Advances in Vestibular Schwannoma Microneurosurgery

Luciano Mastronardi
Takanori Fukushima • Alberto Campione
Editors

Advances in Vestibular Schwannoma Microneurosurgery

Improving Results with New Technologies

 Springer

Editors
Luciano Mastronardi
Unit of Neurosurgery
ASLRoma1
San Filippo Neri Hospital
Roma
Italy

Takanori Fukushima
Neurosurgery Department
Duke Medical Center
Raleigh, NC
USA

Alberto Campione
Unit of Neurosurgery
ASLRoma1
San Filippo Neri Hospital
Rome
Italy

ISBN 978-3-030-03166-4   ISBN 978-3-030-03167-1   (eBook)
https://doi.org/10.1007/978-3-030-03167-1

Library of Congress Control Number: 2018966545

This Springer imprint is published by the registered company Springer Nature Switzerland AG
The registered company address is: Gewerbestrasse 11, 6330 Cham, Switzerland

Foreword

Fortunate are the neurosurgeons who have the opportunity to visit the Department of Neurosurgery at San Filippo Neri Hospital in Rome and Carolina Neuroscience Institute in Raleigh, North Carolina. I am fortunate to have been one of them for so many years.

Visitors to the heads of these departments, authors of this book, just can experience their natural warmth, honesty, and spontaneous integrity together with a gentle sense of humor and love for music, together with an intense devotion for perfecting their surgical experiences.

For me, it is an honor to be invited to write this foreword as their lifetime close friend, having the same basic philosophy. This philosophy is simple and clean. I am thankful for our continued long-term professional cooperation and friendship, which feels like brothership.

For more than 20 years, I cooperated with the authors, my lifetime mentors, heroes, and friends.

Both are recognized as world-class neurosurgeons.

The book is based upon the rich personal experiences of both neurosurgeons . That other neurosurgeons may differ with some of these principles is to be expected, and "to each his own."

The stepwise illustrations and text bring an understanding of surgical anatomy and important minute details of different surgical approaches and techniques in the best way.

Through the years, they have made it a practice to sketch the steps and surgical anatomy of these operative procedures, improving every time while teaching cadaver hands-on courses.

As the title indicates, Drs. Mastronardi and Fukushima have written about neurosurgery experiences on acoustic schwannomas, as performed in both centers; however, knowing them so well, they have done much more. Both the authors are famous teachers—and teaching is an art. The authors are truly master neurosurgeons, and attention will be paid to every detail, including the operation room and table, their personal habits, and the instruments: "the freedom of art."

Surgery of acoustic neuroma is one of the most difficult in neurosurgery. Surgeons must have extremely precise super-micro-operative technical skills, proper usage of updated super-micro-instruments, sufficient knowledge of displaced C-P angle microanatomy, and ample clinical case experience. These principles will result in surgeries with no mortality and negligible morbidity, including normal facial function and preservation of hearing or usable hearing function.

Readers will find considerable practical advice in each chapter, and especially important points will be summarized in all approaches: retrosigmoid, middle fossa, and translabyrinthine. This book contains some of the most accurate and beautiful illustrations of microsurgery of acoustic schwannomas I have ever seen. It deserves to be presented in the language of every neurosurgeon in the world.

This volume is a symbol of the exhaustive labors of the authors and should also be looked upon as eloquent evidence of the high professional caliber of the neurosurgical program at both neurosurgical centers, both in care of patients and in education of neurosurgeons.

Ghent, Belgium   Luc F. De Waele

Preface

New Technologies Available for the Surgical Treatment of Acoustic Neuromas

The aim of this book is to report the results of microsurgical removal of acoustic neuromas (AN) using new technologies: flexible handheld laser fibers, Sonopet Ultrasound Aspirator, Facial Nerve "detector" monopolar stimulation (to localize position and course of nerve), BAERs with LS CE-Chirp stimulus for hearing preservation, injectable bone substitute for closure, and so on. We report retrospective clinical observations on patients operated on during the last 8 years.

From September 2010 to April 2018, 160 consecutive patients suffering from AN have been operated on with a microsurgical technique by keyhole retrosigmoid (RS) approach in the Division of Neurosurgery of San Filippo Neri Hospital, ASLRoma1. In more than 100 cases 2μ-Thulium laser fiber was used for cutting, vaporizing, and removal of tumor. In the same period, Sonopet Ultrasound aspirator was used for tumor debulking and/or opening of the internal auditory canal in all cases. From May 2015, hearing preservation by means of LS-CE-Chirp BAER was attempted in patients with preoperative socially useful hearing (AAO-HNS class A and B). From December 2017 we started to check the total removal of tumor inside the internal auditory canal (IAC) with the flexible endoscope for completing tumor removal near the fundus.

Overall time from incision to skin suture changed in relation to size of tumor and was not affected by the use of new technologies. Facial nerve function was clinically assessed with the House-Brackmann (HB) scale (I–IV) preoperatively, in the early postoperative period (after 1 week), and at 6-month follow-up. In three cases a preoperative facial nerve palsy was observed (2 HB III and 1 HB IV, respectively). In the remaining cases, facial nerve preservation rate (HB I) was more than 90% at 6 months after surgery. Hearing preservation rate (AAO-HNS A and B, preoperatively and postoperatively) was about 50%. Total tumor and "nearly total" removal was possible in about ¾ of cases and subtotal in about 20%. Dura closure with underlying autologous pericranium and injectable bone substitute for closure after bone flap repositioning minimized postoperative CSF leakage.

In conclusion, the use of new technologies in AN microsurgery appeared to be safe and subjectively seems to facilitate tumor resection, especially in "difficult" conditions (e.g., highly vascularized or hard tumors). The good functional outcome following conventional microsurgery could be further improved and the extent of tumor removal could be increased by the new tools available in the neurosurgical armamentarium.

Rome, Italy   Luciano Mastronardi

Contents

Part I

General Aspects

Introduction: Clinical and Radiological Diagnosis

1

Alberto Campione, Guglielmo Cacciotti,
Raffaelino Roperto, Carlo Giacobbo Scavo, Lori Radcliffe,
and Luciano Mastronardi

Vestibular schwannomas (VSs) (also known as acoustic neuromas) arise from Schwann cells, which form the myelin sheath around the vestibulocochlear nerve.

1.1 Epidemiology

VSs predominantly affect adults in their fifth and sixth decades and are indeed much rarer in children, in which are mainly due to neurofibromatosis type 2.

VSs account for 5–10% of intracranial tumors and are the most common neoplastic lesions in the cerebellopontine angle. The overall incidence is approximately 1 per 100,000 persons per year and appears to be increasing because of longer life expectancy and improved diagnostic tools. For these reasons, the mean tumor size at time of diagnosis has progressively declined over the years and is now 10–15 mm, with larger tumors being only a minority [1–4].

No significant differences in incidence between sexes or prevalence of side have been described in the literature.

1.2 Risk Factors

The main risk factor for VS is the exposure to radiation. Two main circumstances have been associated with a higher incidence of the disease:

- Exposure to high-dose ionizing radiation [5]
- Childhood exposure to low-dose radiation for benign head and neck conditions [6, 7]

A. Campione (✉) · G. Cacciotti · R. Roperto · C. Giacobbo Scavo · L. Mastronardi
Department of Neurosurgery, San Filippo Neri Hospital—ASLRoma1, Rome, Italy
e-mail: mastro@tin.it

L. Radcliffe
Carolina Neuroscience Institute, Raleigh, NC, USA
e-mail: lori@carolinaneuroscience.com

© Springer Nature Switzerland AG 2019
L. Mastronardi et al. (eds.), *Advances in Vestibular Schwannoma Microneurosurgery*, https://doi.org/10.1007/978-3-030-03167-1_1

It has recently been statistically ascertained that the exposure to leisure noise positively correlates with an elevated risk of VS. Indeed, severe acoustic trauma from impulse noise can cause mechanical damage of N VIII and the surrounding tissues. From a biochemical perspective, loud acoustic stimulation induces electrolytes disequilibrium and release of free radicals in cochlear fluids, which could in turn be responsible for DNA damage in cochlear hair cells. Thus, although the exact mechanism is still unclear, it is plausible that VSs may arise due to chronic trauma due to impulse noise [8].

The role of cellular telephone in the pathogenesis of VS remains unclear and controversial.

1.3   Pathogenesis and Pathology

From a genetic perspective, the origin of most sporadic VSs is the biallelic inactivation of gene *NF2*, which encodes a protein called merlin (also known as schwannomin) that acts as a tumor suppressor. The gene *NF2* was first discovered as the locus on chromosome 22 harboring the mutation responsible for familiar and bilateral VSs seen in neurofibromatosis type 2 (NF2) [1, 2].

VSs are classified as WHO grade I tumors and have a locally compressing effect rather than an infiltrative tendency. The growth rate is <1 mm per year for more than 60% of patients and >3 mm per year for 12%. The average ki67 (MIB-1) index ranges between 1.86 and 1.99%. However, some differences in terms of ki67 index have been reported between unilateral and bilateral VSs, as well as between growing and stable ones [9–11]. The prognostic value of such discrepancies has been deemed uncertain and in need of future study [12]. Malignant degeneration is exceedingly rare.

Macroscopically, VSs appear as pale, capsulated globoid masses displacing or splaying surrounding neural structures. Occasional foci of hemorrhage or cystic degeneration may also be observed.

The site of origin is the inferior vestibular nerve (IVN) in 70% of cases and the superior vestibular nerve (SVN) in 20% of cases. In both the circumstances, VSs frequently arise at the Obersteiner-Redlich junction, i.e., the point of encounter of central and peripheral myelin, near to porus acusticus. Less frequently, the tumor can arise close to the meatus, and in this case it is referred to as the "medial variety," with a small amount of tumor in the lateral part of the internal auditory canal [13]. In rarer cases—10% of total—VSs may arise from the cochlear nerve [14].

As for the displacement of cranial nerves, intraoperative stimulation and neuromonitoring have enabled to trace the course of both facial (N VII) and cochlear (N VIII) nerves. In 70% of cases, the position of N VII is ventral or ventral-superior to the tumor. Alternatively, N VII may be displaced superiorly (20% of cases), inferiorly or ventral-inferiorly (10%), and, rather exceptionally, dorsally. The position of N VIII is usually inferior or ventral-inferior [14, 15].

Microscopically, neoplastic Schwann cells appear as arranged in two different tissue patterns: Antoni A (dense cellularity) and Antoni B (sparse cellularity). As VSs induce angiogenesis, this may result in telangiectatic formations and subsequent intratumoral hemorrhages. A thick collagenous capsule is usually present [1, 4].

1.4   Clinical Presentation

The clinical presentation of VSs correlates with the structures gradually and chronically compressed by the tumor. These are in the first place N VIII, trigeminal nerve (N V), and N VII. Later, as the tumor enlarges, lower cranial nerves may be involved and then the cerebellum and the brainstem.

- The most common symptoms are hearing loss (95% of patients) and tinnitus (63%), usually with a chronic onset—although acute cases have been occasionally reported. Preoperative tinnitus has been reported as a negative prognostic sign for hearing preservation [16].
- Disequilibrium (61%) is typically mild to moderate in intensity and fluctuating in timing.
- The involvement of N V (17%) is often reported after the hearing loss and mainly presents as facial paresthesia, hypoesthesia, or pain.
- Facial paresis (6%) is usually of chronic onset and correlates (in most of cases) with tumors large enough to compress the intralabyrinthine tract of N V or even the geniculate ganglion. However, recent reports have described cases of acute facial paresis as the exording sign of still intracanalicular VSs (Mastronardi et al., in publication). This peculiar presentation is exceptional (1%) and may pose a challenging differential diagnosis with N VII schwannoma, meningioma, cavernoma, or malignancy.
- The symptoms of tumor progression may be of different intensity and timing. Headache (32%) is rather common and early. Instead, nausea and vomiting (9%) correlate with compression of cerebellum and brainstem, which may lead to hydrocephalus.

1.5   Diagnosis

The diagnosis of VS requires both clinical suspicion and laboratory/imaging tests.

VS must be suspected in case of unilateral hearing loss of chronic onset, with a positive Rinne test and a Weber test lateralized on the unaffected side. Both the tests are needed to confirm the sensorineural origin of the hearing impairment.

Functional laboratory tests constitute the next step of diagnostic workup. Audiometry is the best initial screening test as only 5% of affected patients have normal results. Pure tone and speech discrimination audiometry should be performed for a correct classification of the patient's hearing damage. Test results typically show hearing loss for high frequencies and a disproportionately negative speech discrimination score [2, 4]. The American Academy of Otolaryngology-Head and Neck Surgery (AAO-HNS) hearing classification is the current grading system that, according to pure-tone average and speech discrimination percentage, determines whether the hearing is still functional and to what extent (Table 1.1).

Auditory brainstem evoked response (ABR) is a further functional screening measure that may be used in case of abnormal results of audiometric tests. The main

Table 1.1 American Academy of Otolaryngology-Head and Neck Surgery (AAO-HNS) hearing classification

Class	Pure-tone average (dB)	Speech discrimination (%)
A: Useful	≤30	≥70
B: Socially useful	>30 and ≤50	≥50
C: Capable of aid (serviceable)	>50	≥50
D: Nonfunctional	Any level	<50

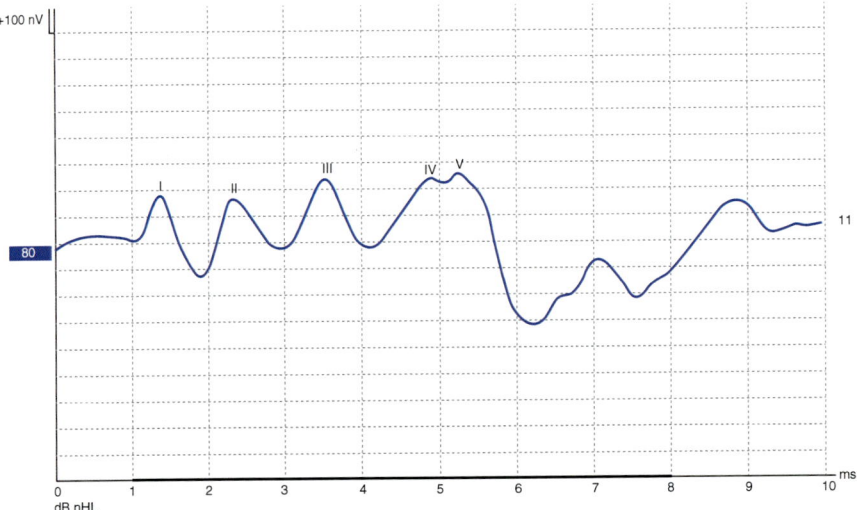

Fig. 1.1 Normal ABR. *Reprinted from Clinical Neurology and Neurosurgery, 165, Luciano Mastronardi, Ettore Di Scipio, Guglielmo Cacciotti, Raffaelino Roperto, Vestibular schwannoma and hearing preservation: Usefulness of level specific CE-Chirp ABR monitoring. A retrospective study on 25 cases with preoperative socially useful hearing, Pages No. 108–115, 2018, with permission from Elsevier*

advantage of ABR is that it is a patient-independent test and shows objective results about the damage to the different levels of the auditory pathway. ABR results typically show a major delay in cochlear nerve conduction, with increased latency in wave III and, later, in waves V and VI [4] (Figs. 1.1 and 1.2).

The technique of ABR is not only useful as a diagnostic tool but also as an intra-operative neuro-monitoring instrument that enables the surgeon to correctly identify and preserve N VIII. At the end of surgical operation, ABR reflects the functional state of N VIII, which is shown in Fig. 1.3. ABR may appear as normal, destructured, or delayed.

Imaging tests constitute the final step of diagnostic workup. MRI with gadolinium contrast is the gold standard for radiological diagnosis. VS appears as lesions of the internal auditory canal (IAC) with variable extension into the cerebellopontine angle (CPA). Different grading systems have been proposed to stage tumor progression. Samii's classification is one of the most commonly used and is mainly based on the anatomical relationship around the tumor [17] (Table 1.2) (Figs. 1.4 and 1.5).

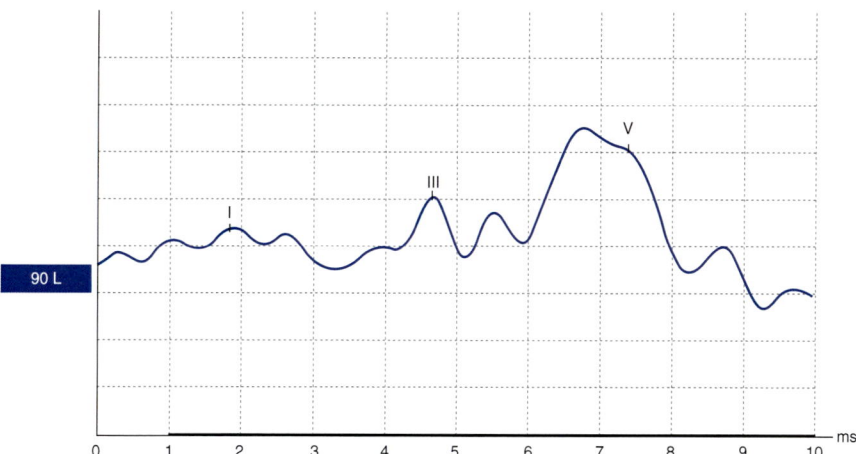

Fig. 1.2 Pathologic ABR. *Reprinted from Clinical Neurology and Neurosurgery, 165, Luciano Mastronardi, Ettore Di Scipio, Guglielmo Cacciotti, Raffaelino Roperto, Vestibular schwannoma and hearing preservation: Usefulness of level specific CE-Chirp ABR monitoring. A retrospective study on 25 cases with preoperative socially useful hearing, Pages No. 108–115, 2018, with permission from Elsevier*

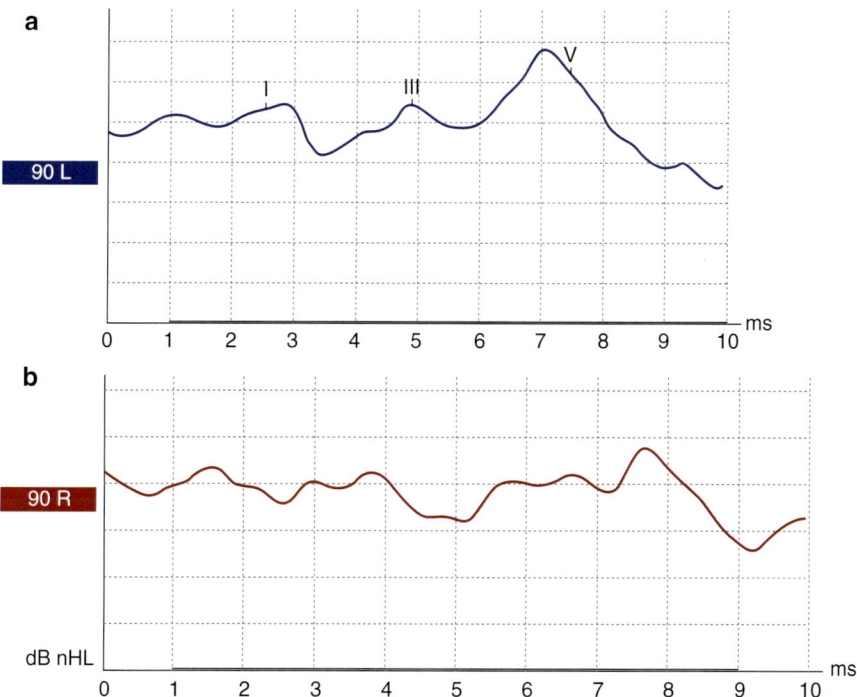

Fig. 1.3 Postoperative ABRs. (**a**) Normal postoperative ABR (registered from the same patient as in Fig. 1.2). (**b**) Destructured ABR. (**c**) Delayed ABR. *Reprinted from Clinical Neurology and Neurosurgery, 165, Luciano Mastronardi, Ettore Di Scipio, Guglielmo Cacciotti, Raffaelino Roperto, Vestibular schwannoma and hearing preservation: Usefulness of level specific CE-Chirp ABR monitoring. A retrospective study on 25 cases with preoperative socially useful hearing, Pages No. 108–115, 2018, with permission from Elsevier*

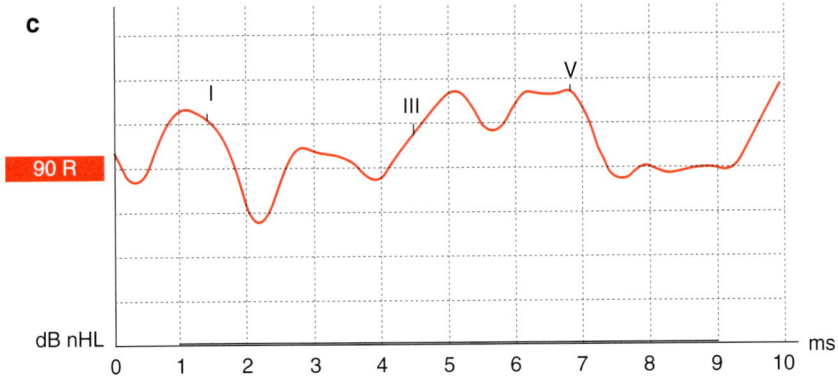

Fig. 1.3 (continued)

Table 1.2 Samii's classification of VSs

	Tumor description
T1	Confining to IAC
T2	Surpassing IAC
T3a	Tumor occupying CPA
T3b	Tumor occupying CPA and contacting brainstem without compression
T4a	Tumor compressing the brainstem
T4b	Severe brainstem displacement and deformation of the fourth ventricle under tumor compression

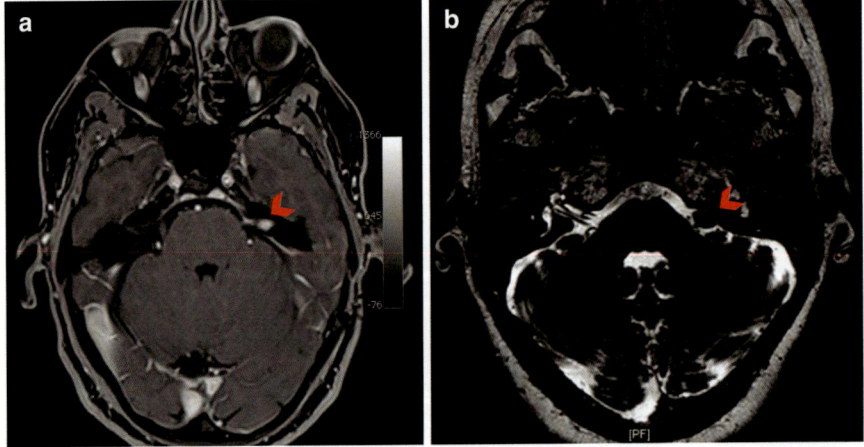

Fig. 1.4 Samii's classification of VSs. (**a**) Stage T1, tumor indicated by arrow. (**b**) Stage T2, tumor indicated by arrow

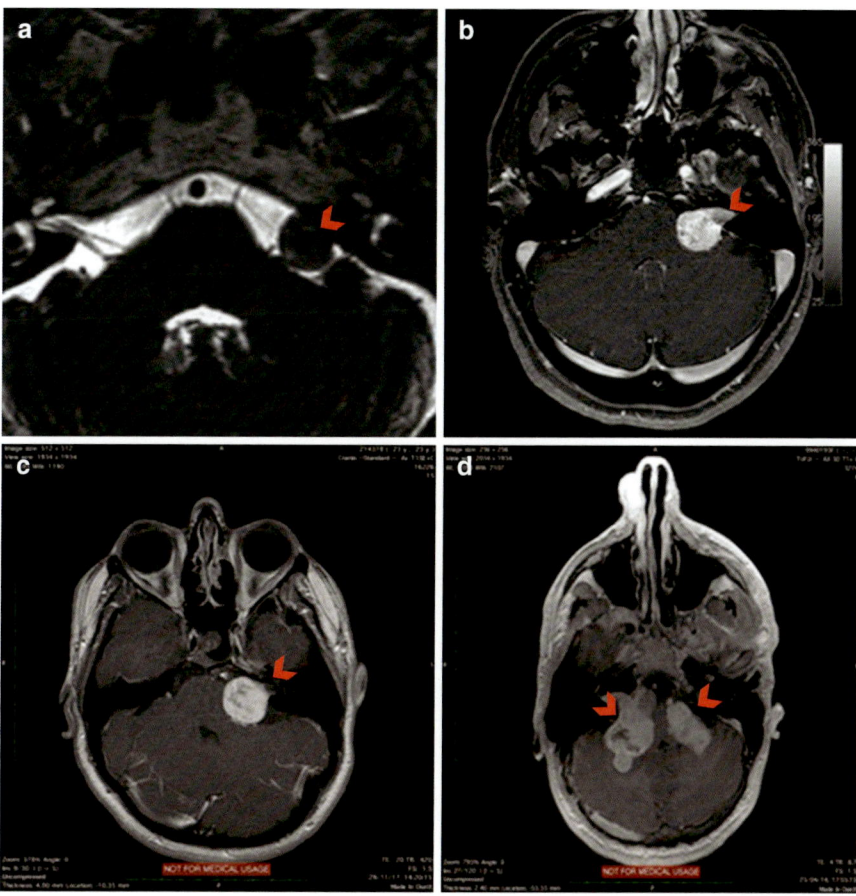

Fig. 1.5 Samii's classification of VSs. (**a**) Stage T3a, tumor indicated by arrow. (**b**) Stage T3b, tumor indicated by arrow. (**c**) Stage T4a, tumor indicated by arrow. Note that despite the brainstem compression the fourth ventricle still retains its normal morphology. (**d**) Stage T4b, tumors indicated by arrows. Note that this is a rare case of bilateral giant VSs. Figures (**b** and **c**) *reprinted from Clinical Neurology and Neurosurgery, 165, Luciano Mastronardi, Ettore Di Scipio, Guglielmo Cacciotti, Raffaelino Roperto, Vestibular schwannoma and hearing preservation: Usefulness of level specific CE-Chirp ABR monitoring. A retrospective study on 25 cases with preoperative socially useful hearing, Pages No. 108–115, 2018, with permission from Elsevier*

VSs characteristically have an intracanalicular component which widens the porus acusticus, leading to the "trumpeted internal acoustic meatus sign" in MRI. The extracanalicular component typically has a globoid shape and is obviously connected to the cone-like intracanalicular component, thus resembling an "ice-cream cone appearance." VSs are described as isointense/hypointense lesions on T1-weighted images and are strongly contrast enhanced. On T2-weighted images, they are hyperintense. Cystic lesions may be observed in 10–15% of cases, especially when the neoplastic mass reaches a big size.

CT in VS diagnosis is reserved to patients who do not tolerate MRI but may be useful—especially in the case of giant tumors—to observe the degree of bone erosion around the tumor for a better operative planning.

Radiological differential diagnosis of VSs includes meningiomas and epidermoid cysts. Meningiomas have a similar appearance on T1- and T2-weighted images. However, calcifications are usually present inside the neoplastic mass, and a broad dural base may also be observed. In addition, meningiomas may induce hyperostosis of adjacent bone, while VSs may be associated with bone erosion. Exceptionally, meningiomas can grow inside the IAC [18, 19], and in these cases, the differential diagnosis can be obtained only during surgical removal [20–22]. Epidermoid cysts are isointense to cerebrospinal fluid in both T1- and T2-weighted images, are not contrast enhancing, and do not extend in the IAC.

References

1. Di Ieva A, Lee JM, Cusimano MD. Handbook of skull base surgery. New York: Thieme; 2016. xxvii, 978 p.
2. Park JK, Vernick DM, Ramakrishna N. Vestibular schwannoma (acoustic neuroma). In: Post TW, editor. UpToDate. Waltham: UpToDate. Accessed 14 Jan 2018.
3. Quiñones-Hinojosa A, Rincon-Torroella J. Video atlas of neurosurgery: contemporary tumor and skull base surgery. 1st ed. New York: Elsevier; 2017. xxx, 285 p.
4. Winn HR. Youmans and Winn neurological surgery. 7th ed. Philadelphia, PA: Elsevier; 2017.
5. Sughrue ME, Yeung AH, Rutkowski MJ, Cheung SW, Parsa AT. Molecular biology of familial and sporadic vestibular schwannomas: implications for novel therapeutics. J Neurosurg. 2011;114(2):359–66.
6. Schneider AB, Ron E, Lubin J, Stovall M, Shore-Freedman E, Tolentino J, et al. Acoustic neuromas following childhood radiation treatment for benign conditions of the head and neck. Neuro Oncol. 2008;10(1):73–8.
7. Shore-Freedman E, Abrahams C, Recant W, Schneider AB. Neurilemomas and salivary gland tumors of the head and neck following childhood irradiation. Cancer. 1983;51(12):2159–63.
8. Chen M, Fan Z, Zheng X, Cao F, Wang L. Risk factors of acoustic neuroma: systematic review and meta-analysis. Yonsei Med J. 2016;57(3):776–83.
9. Niemczyk K, Vaneecloo FM, Lecomte MH, Lejeune JP, Lemaitre L, Skarzyński H, et al. Correlation between Ki-67 index and some clinical aspects of acoustic neuromas (vestibular schwannomas). Otolaryngol Head Neck Surg. 2000;123(6):779–83.
10. Saito K, Kato M, Susaki N, Nagatani T, Nagasaka T, Yoshida J. Expression of Ki-67 antigen and vascular endothelial growth factor in sporadic and neurofibromatosis type 2-associated schwannomas. Clin Neuropathol. 2003;22(1):30–4.
11. Steinhart H, Triebswetter F, Wolf S, Gress H, Bohlender J, Iro H. Growth of sporadic vestibular schwannomas correlates with Ki-67 proliferation index. Laryngorhinootologie. 2003;82(5):318–21.
12. Sughrue ME, Fung KM, Van Gompel JJ, Peterson JEG, Olson JJ. Congress of neurological surgeons systematic review and evidence-based guidelines on pathological methods and prognostic factors in vestibular schwannomas. Neurosurgery. 2018;82(2):E47–8.
13. Dunn IF, Bi WL, Erkmen K, Kadri PA, Hasan D, Tang CT, et al. Medial acoustic neuromas: clinical and surgical implications. J Neurosurg. 2014;120(5):1095–104.
14. Mastronardi L, Cacciotti G, Roperto R, Di Scipio E, Tonelli MP, Carpineta E. Position and course of facial nerve and postoperative facial nerve results in vestibular schwannoma microsurgery. World Neurosurg. 2016;94:174–80.

15. Sameshima T, Morita A, Tanikawa R, Fukushima T, Friedman AH, Zenga F, et al. Evaluation of variation in the course of the facial nerve, nerve adhesion to tumors, and postoperative facial palsy in acoustic neuroma. J Neurol Surg B Skull Base. 2013;74(1):39–43.
16. Mastronardi L, Cacciotti G, Roperto R, DI Scipio E. Negative influence of preoperative tinnitus on hearing preservation in vestibular schwannoma surgery. J Neurosurg Sci. 2017. https://doi.org/10.23736/S0390-5616.17.04187-X.
17. Wu H, Zhang L, Han D, Mao Y, Yang J, Wang Z, et al. Summary and consensus in 7th International Conference on acoustic neuroma: an update for the management of sporadic acoustic neuromas. World J Otorhinolaryngol Head Neck Surg. 2016;2(4):234–9.
18. Amato MC, Colli BO, Carlotti Junior CG, dos Santos AC, Féres MC, Neder L. Meningioma of the internal auditory canal: case report. Arq Neuropsiquiatr. 2003;61(3A):659–62.
19. Watanabe K, Cobb MIH, Zomorodi AR, Cunningham CD, Nonaka Y, Satoh S, et al. Rare lesions of the internal auditory canal. World Neurosurg. 2017;99:200–9.
20. Asaoka K, Barrs DM, Sampson JH, McElveen JT, Tucci DL, Fukushima T. Intracanalicular meningioma mimicking vestibular schwannoma. AJNR Am J Neuroradiol. 2002;23(9):1493–6.
21. Chae SW, Park MK. Meningioma mimicking vestibular schwannoma. Ear Nose Throat J. 2011;90(7):299–300.
22. Roos DE, Patel SG, Potter AE, Zacest AC. When is an acoustic neuroma not an acoustic neuroma? Pitfalls for radiosurgeons. J Med Imaging Radiat Oncol. 2015;59(4):474–9.

Treatment Options and Surgical Indications

2

Luciano Mastronardi, Alberto Campione,
Raffaelino Roperto, Albert Sufianov,
and Takanori Fukushima

The management of vestibular schwannomas (VSs) is diverse and depends on characteristics of both the patient and the tumor. Tumor size and growth pattern as well as patient's age, symptoms, and comorbidities determine the treatment of choice among three main options: conservative therapy with watchful waiting, radiation therapy (RT), and surgery.

The goal of modern management of VSs is to improve the quality of life and to preserve the neurological functions while maintaining mortality and morbidity rates as low as possible. The treatment strategy must also be weighed against the patient's desires and motivation so that the final approach is carefully individualized [1–3].

2.1 "Wait and Watch" (Fig. 2.1)

Watchful waiting consists of periodical MRI scans to monitor the size of the tumor. Although different growth patterns can be observed in time [4, 5], the natural history of VSs may follow three different paths: no growth, slow growth (max 2 mm[1]/year), and fast growth (≥3 mm/year). The scheme of scans to follow is not standard-

---

[1] As the thickness of the slices in MRI scans is 1 mm, anything smaller is attributable to interobserver variation; this is the reason why the cutoffs to discriminate between slow and fast growth are integers.

L. Mastronardi (✉) · A. Campione · R. Roperto
Department of Neurosurgery, San Filippo Neri Hospital—ASLRoma1, Rome, Italy
e-mail: mastro@tin.it

A. Sufianov
Federal Centre of Neurosurgery, Tyumen, Russia

T. Fukushima
Division of Neurosurgery, Duke University Medical Center, Carolina Neuroscience Institute, Raleigh, NC, USA
e-mail: Fukushima@carolinaneuroscience.com

© Springer Nature Switzerland AG 2019
L. Mastronardi et al. (eds.), *Advances in Vestibular Schwannoma Microneurosurgery*, https://doi.org/10.1007/978-3-030-03167-1_2

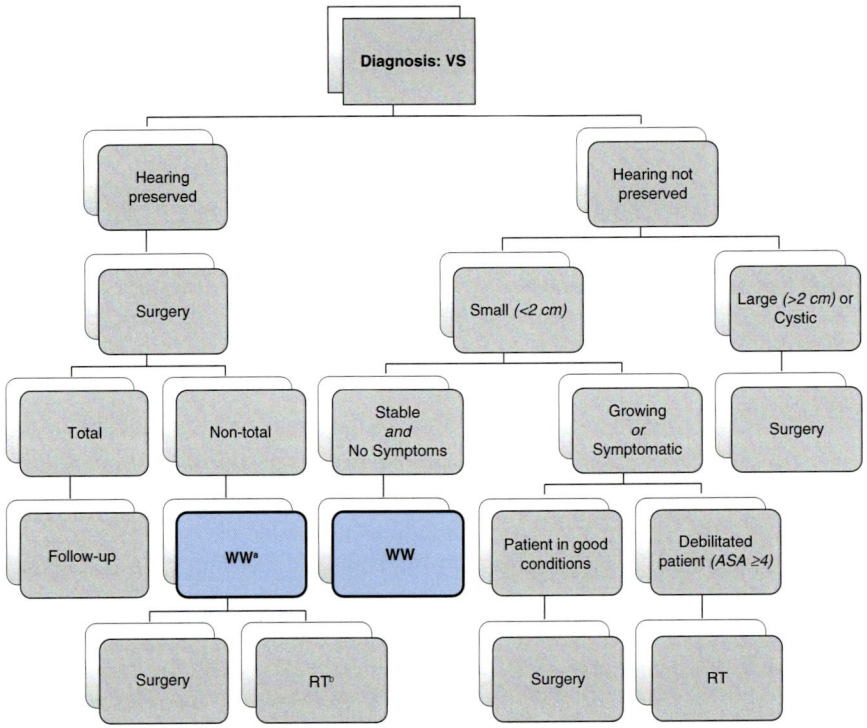

Fig. 2.1 Treatment strategies in vestibular schwannomas (VSs). The paths leading to watchful waiting are emphasized so as to clarify its main indications. ªWatchful waiting. ᵇRadiation therapy

ized, and diverse ones have been proposed in different studies [4–7]. As a general rule, annual imaging studies should be performed for the rest of the patient's life. On one hand, it has been argued that as many as 64% of VSs showed a growth pattern that had remained uniform for the first 5 years of follow-up, thus confirming a large predictability of the natural history of those tumors [5, 6]. Nevertheless, cases of sudden tumor growth after over 10 years of follow-up have been reported [5] and would have been missed in case of a time-limited watchful waiting.

This type of strategy mainly applies to incidentally diagnosed VSs with the following characteristics:

- Patient with nonfunctional hearing (AAO-HNS class C/D)
- Tumor size ≤2 cm at diagnosis [1, 4]
- Tumor growth rate max 2 mm/year
- No neurological symptoms due to trigeminal (NV), facial (NVII), or vestibulocochlear (NVIII) nerve compression or hydrocephalus

As far as hearing function in concerned, some authors have proposed to extend the watchful waiting strategy to intrameatal VSs in younger patient who still have serviceable hearing as these tumors show a higher tendency to remain stable over time when compared to larger ones [5]. However, as the goal in the treatment of VSs is to preserve neurological functions and, in addition, hearing deterioration can occur even in non-growing tumors, small VSs in younger patients who still retain hearing function should be treated with hearing preservation surgery [8].

As for the patient's characteristics that have been proposed to encourage a watchful waiting approach—being tumor size and growth rate constant and within the aforementioned values, these are [1–3]:

- Age > 65 years
- Poor clinical conditions contraindicating any interventional strategy

Only a minority of patients eligible for watchful waiting still retain hearing functions at the time of diagnosis. This may be due to the concurrent setting of presbycusis as they are mostly elderly. In terms of hearing preservation, 30–50% [6, 9] of patients maintain hearing function during a 5-year watchful waiting follow-up.

The greatest concern about watchful waiting is the risk of injury to NVII. According to recent studies [4, 5], a conservative strategy does not imply per se a poor facial outcome. However, the possible increase of tumor size may lead to poorer results in NVII preservation in case of surgical resection. This risk should be weighed against the fact that, when the cohort is chosen accurately, watchful waiting fails in a minority of cases, as reported in the series by the same authors [4, 5].

2.2 Radiation Therapy (Fig. 2.2)

RT protocols for the management of VSs differ in dose and radiation/particles administered. There are three types of RT: stereotactic radiosurgery (SRS or Gamma Knife), fractionated stereotactic radiation therapy (FSRT), and proton therapy (PT). The common feature of these therapies is local control rather than tumor eradication [1, 2, 10]. No studies have compared two or all the three modalities [11].

RT is indicated in case of [1, 2]:

- Patient with nonfunctional hearing (AAO-HNS class C/D)
- Tumor size <2 cm
- Tumor growth is ≥3 mm/year
- Poor clinical conditions contraindicating surgical intervention (ASA ≥ 4)

Older or comorbid patients may benefit from RT instead of surgery so as to avoid the complications related to anesthesia and surgery itself (Fig. 2.2). However, RT itself may lead to complications such as hearing loss, NV and NVII deficits, adhesions, hydrocephalus, and potential (although very rare) malignant transformation [1, 11]. Rare sequelae yet to be considered are delayed severe headache, severe facial pain, and new motor deficits.

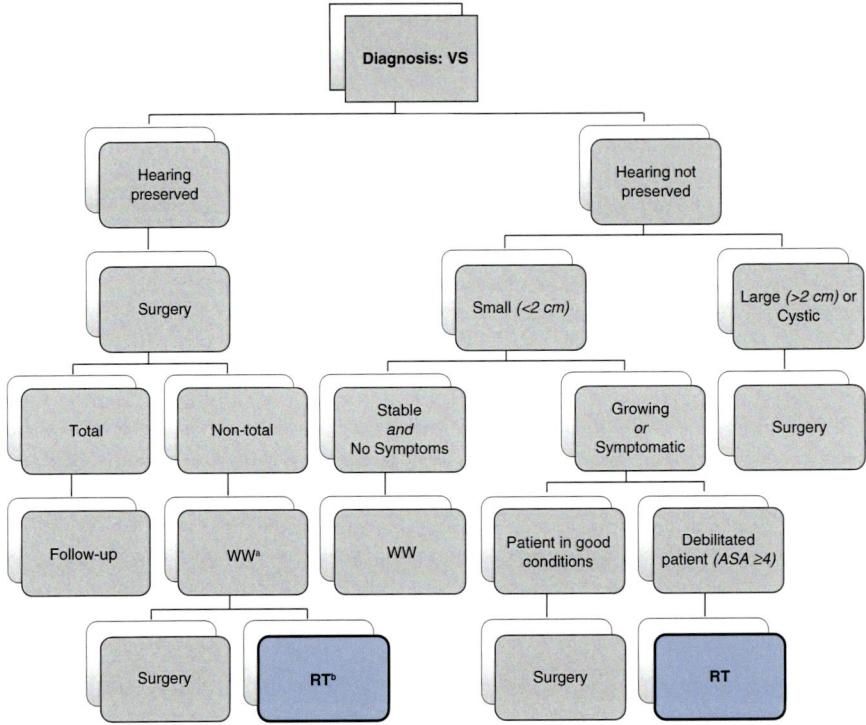

Fig. 2.2 Treatment strategies in vestibular schwannomas (VSs). The paths leading to radiation therapy are emphasized so as to clarify its main indications. [a]Watchful waiting. [b]Radiation therapy. After "non-total" removal of tumor, retreatment could be necessary if the residue grows during WW follow-up

Moreover, the treatment of debilitated (ASA $\geq$ 4) patients still remains controversial. While for larger tumors a debulking surgery, i.e., intended not to be total, may be the sole possible treatment strategy, small (<2 cm) tumors and—even more—intrameatal ones may be eligible for watchful waiting. Nevertheless, many studies have proposed RT for this kind of VSs as well, independently of the patient's hearing function. Recent systematic reviews [4, 5] have examined the role of RT in the treatment of VSs and found that it is neither standardized nor immediately clear. In fact, as a considerable portion of these tumors would remain stable or grow slowly, the successes claimed of RT may have been at least partially facilitated by the nature of the tumor itself. As a consequence, recent guidelines recommend that intrameatal and small VSs without tinnitus be observed as observation does not have a negative impact on tumor growth or hearing preservation compared to RT treatment [11].

RT has an additional role in retreatment in case of failure of previous surgery or RT protocol [1, 9–11]. However, in the perspective of multiple recurrences that require surgical intervention, RT may pose problems in terms of NVII preservation because of the numerous radiation-induced adhesions.

Hearing preservation rate after RT is approximately 50% with a strong variability among different studies [1, 6, 12].

Trigeminal (NV) and facial nerve (NVII) preservation rate after RT is >95% and has been reported as higher than after surgery [1].

2.3   **Surgery** (Fig. 2.3)

Surgical resection is still the sole treatment modality that—if total—guarantees tumor eradication. As far as the extent of surgical resection is concerned, specific definitions have been elaborated so that surgery can be defined as:

- Total, if 100% of tumor is resected.
- Nearly total, if a millimetric (<7 mm) tumor capsule residue is left along NVII or on the brainstem and 95–99% of tumor is resected.
- Subtotal, if 90–95% of tumor is resected.
- Partial, if <90% of tumor is resected.

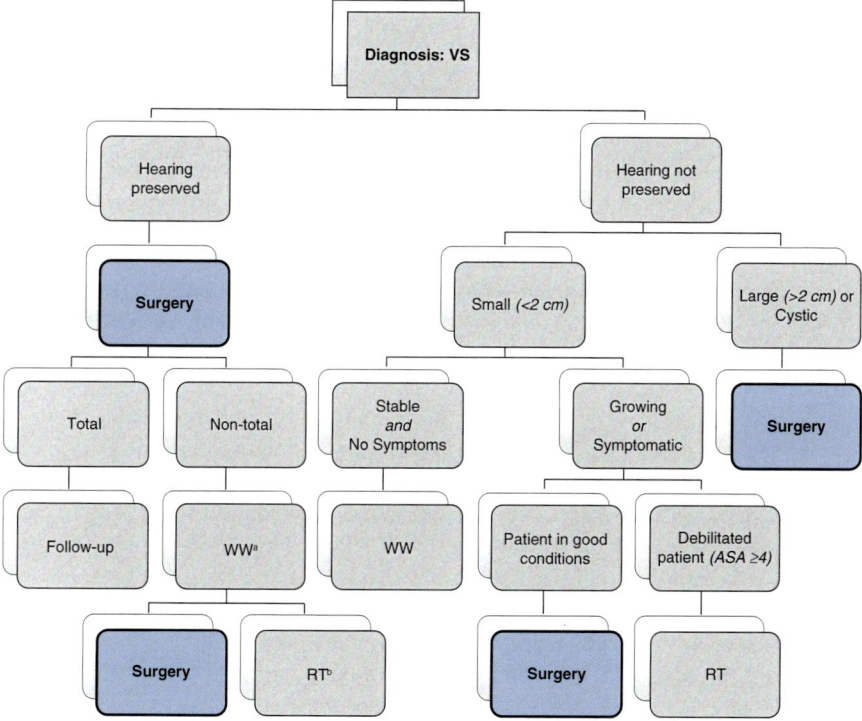

Fig. 2.3 Treatment strategies in vestibular schwannomas (VSs). The paths leading to surgery are emphasized so as to clarify the main indications for microsurgical resection. Despite it is explicitly shown only once in figure, the algorithm of total/non-total surgery is valid and must be taken into consideration every time surgery is performed. [a]Watchful waiting. [b]Radiation therapy

- Debulking, if the purpose of surgery is not to eradicate the tumor but to decompress the nervous tissue—debilitated patients with large tumors would benefit from this strategy.

Three surgical approaches have been developed for the management of VSs:

- Retrosigmoid approach: Used in the majority of cases, it is indicated for small and large tumors and when hearing preservation is a goal of the operation.
- Translabyrinthine approach: It is indicated for deaf patients with larger tumors.
- Middle fossa approach: Currently less used, it is indicated for small intracanalicular tumors when hearing preservation is a goal of the operation.

Surgery is indicated in case of [1–3]:

- Patient presenting socially useful hearing (AAO-HNS class A/B) at diagnosis.
- Tumor size >2 cm, with the exception of smaller tumors, including the intracanalicular ones, in patients with preoperative socially useful hearing [8, 9].
- Tumor growth >2 mm per year.
- Presence of neurological symptoms: in the rare case of VS-induced trigeminal neuralgia, surgery is preferred to SRS for the management and resolution of the symptom [8].
- Cystic degeneration of tumor on MRI [1].

Patients <65 years and in good clinical conditions are ideal candidates for the operation.

Complete tumor removal is feasible in almost all the cases. However, when hearing preservation is attempted and in order to avoid any damage to NV or NVII, the resection may not be completed—this is the case in approximately 15% of all VS surgeries [2]. Tumor remnants after partial or subtotal removal have been traditionally treated with adjuvant RT to lower the risk of recurrence [8], which is reported in the literature as high as 27.6% at 2 years after an incomplete resection. Preoperative tumor volume directly correlates with postoperative remnant volume, and both the parameters are predictive factors for a recurrence, defined as remnant growth seen in MRI scans during follow-up. For larger tumors (Koos grade IV VSs, equivalent to T4a/T4b tumors in Samii's classification), one would expect that aggressive surgical strategy would be the best way to avoid recurrences. However, new perspectives are emerging, and a recent study [13] has reported a recurrence rate of 16% after intentional near-total resection with "wait-and-watch" approach of the tumor remnant and RT or second surgery in case of remnant regrowth.

Hearing preservation after surgery is reported in almost 66% of patients [6], which is the highest rate among the three available treatment modalities for VSs. Hearing preservation surgery is particularly important in those patients who have AAO-HNS class A/B hearing but already show initial hearing loss—a predictive factor of further hearing loss over time. In this cohort, surgery represents the best strategy to preserve the patient's neurological functions.

Facial nerve (NVII) preservation after surgery strongly depends on—and is inversely correlated to—two factors: larger tumor size and previous RT [8]. In fact, when microsurgical resection is necessary after RT, there is an increased likelihood of subtotal resection and decreased NVII function [8]. Surgery may be required after RT in case of multiple recurrences that still remain a controversial subject. A recent retrospective analysis [14] on recurrent VSs examined the early NVII outcome in patients who received a second treatment after failing of primary surgery. Two groups were studied: the first had undergone adjuvant RT after primary surgery, while the other had not received it. No differences in the outcome were observed between the two groups, but in case of secondary treatment, a previous adjuvant RT was associated with more complications. In fact, RT for recurrent VSs caused delayed NVII palsy in 5–21% of cases, and the effect was more evident in those patients who underwent adjuvant RT. On the other hand, secondary surgery after adjuvant RT may result in more difficult and incomplete resection. As a consequence, watchful waiting and possible further microsurgical resection are suggested as the best treatment for tumor remnants.

References

1. Di Ieva A, Lee JM, Cusimano MD. Handbook of skull base surgery. New York: Thieme; 2016. xxvii, 978 p.
2. Park JK, Vernick DM, Ramakrishna N. Vestibular schwannoma (acoustic neuroma). In: Post TW, editor. UpToDate. Waltham: UpToDate. Accessed 14 Jan 2018.
3. Quiñones-Hinojosa A, Rincon-Torroella J. Video atlas of neurosurgery: contemporary tumor and skull base surgery. 1st ed. Edinburgh; New York: Elsevier; 2017. p. 285.
4. Patnaik U, Prasad SC, Tutar H, Giannuzzi AL, Russo A, Sanna M. The long-term outcomes of wait-and-scan and the role of radiotherapy in the management of vestibular schwannomas. Otol Neurotol. 2015;36(4):638–46.
5. Prasad SC, Patnaik U, Grinblat G, Giannuzzi A, Piccirillo E, Taibah A, et al. Decision making in the wait-and-scan approach for vestibular schwannomas: is there a price to pay in terms of hearing, facial nerve, and overall outcomes? Neurosurgery. 2018;83(5):858–70.
6. Hoa M, Drazin D, Hanna G, Schwartz MS, Lekovic GP. The approach to the patient with incidentally diagnosed vestibular schwannoma. Neurosurg Focus. 2012;33(3):E2.
7. Nuseir A, Sequino G, De Donato G, Taibah A, Sanna M. Surgical management of vestibular schwannoma in elderly patients. Eur Arch Otorhinolaryngol. 2012;269(1):17–23.
8. Hadjipanayis CG, Carlson ML, Link MJ, Rayan TA, Parish J, Atkins T, et al. Congress of Neurological Surgeons systematic review and evidence-based guidelines on surgical resection for the treatment of patients with vestibular schwannomas. Neurosurgery. 2018;82(2):E40–3.
9. Wu H, Zhang L, Han D, Mao Y, Yang J, Wang Z, et al. Summary and consensus in 7th International Conference on acoustic neuroma: an update for the management of sporadic acoustic neuromas. World J Otorhinolaryngol Head Neck Surg. 2016;2(4):234–9.
10. Winn HR. Youmans and Winn neurological surgery. 7th ed. Philadelphia: Elsevier; 2017.
11. Germano IM, Sheehan J, Parish J, Atkins T, Asher A, Hadjipanayis CG, et al. Congress of Neurological Surgeons systematic review and evidence-based guidelines on the role of radiosurgery and radiation therapy in the management of patients with vestibular schwannomas. Neurosurgery. 2018;82(2):E49–51.
12. Kondziolka D, Mousavi SH, Kano H, Flickinger JC, Lunsford LD. The newly diagnosed vestibular schwannoma: radiosurgery, resection, or observation? Neurosurg Focus. 2012;33(3):E8.

13. Zumofen DW, Guffi T, Epple C, Westermann B, Krähenbühl AK, Zabka S, et al. Intended near-total removal of Koos grade IV vestibular schwannomas: reconsidering the treatment paradigm. Neurosurgery. 2018;82(2):202–10.
14. Samii M, Metwali H, Gerganov V. Microsurgical management of vestibular schwannoma after failed previous surgery. J Neurosurg. 2016;125(5):1198–203.

Part II

Illustrated Surgical Technique (Step-by-Step)

Patient Positioning

3

Luciano Mastronardi, Alberto Campione,
Guglielmo Cacciotti, Raffaelino Roperto, Fabio Crescenzi,
Ali Zomorodi, and Takanori Fukushima

3.1   Lateral Position [1, 2]

The lateral position, also known as Fukushima position (Fig. 3.1), allows access to the cerebellopontine angle (CPA) and combines satisfactory surgical exposure with anesthesiological safety.

Following intubation, the patient is placed in the lateral decubitus position, the dependent side being contralateral to the lesion. This reduces the neck rotation necessary for a complete surgical exposure and preserves the venous outflow, especially through the contralateral jugular vein. The back of the operating table is then raised at an angle of 10–15° to the floor, and the whole operating table is set to the reverse Trendelenburg position at an angle of 10–15°, allowing less venous congestion in case of a prolonged operation. The patient's back is brought close to the edge of the operating table, and the shoulders are positioned at the cephalad end of the table.

The dependent leg is flexed at approximately 90° at both the hip and the knee, while the nondependent leg is only slightly flexed (Fig. 3.1a).

To prevent pressure injuries, multiple pads are positioned. The heel and the ankle of the dependent leg are padded to relieve the pressure on the peroneal nerve as it passes across fibular head. Two pillows are placed transversely between the knees and one pillow longitudinally between the legs. A gel pad is placed underneath the dependent hip to pad the trochanter. Next, a padded rest is positioned over the gluteal muscles to prevent the patient from rolling backwards—care must be taken to

L. Mastronardi (✉) · A. Campione · G. Cacciotti · R. Roperto · F. Crescenzi
Department of Neurosurgery, San Filippo Neri Hospital—ASLRoma1, Rome, Italy
e-mail: mastro@tin.it

A. Zomorodi · T. Fukushima
Division of Neurosurgery, Duke University Medical Center, Carolina Neuroscience Institute, Raleigh, NC, USA
e-mail: ali.zomorodi@duke.edu; Fukushima@carolinaneuroscience.com

© Springer Nature Switzerland AG 2019
L. Mastronardi et al. (eds.), *Advances in Vestibular Schwannoma Microneurosurgery*, https://doi.org/10.1007/978-3-030-03167-1_3

23

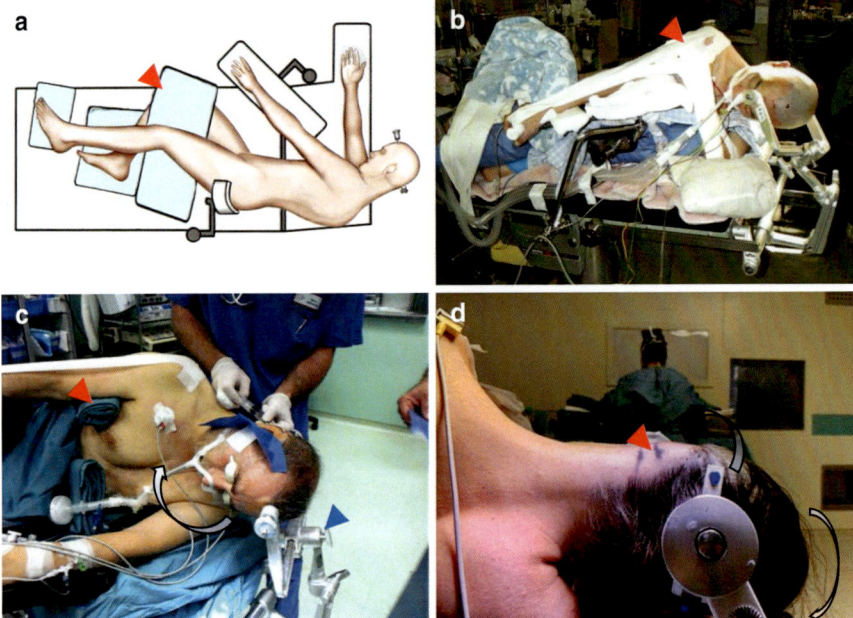

Fig. 3.1 Patient positioning. (**a**) Diagram of a patient in Fukushima position viewed from above. Note the lateral decubitus position with the dependent leg (red arrowhead) flexed at 90° and the nondependent arm (blue arrowhead) placed at 45°. (**b**) Patient in Fukushima position. Note that the nondependent shoulder (arrowhead) has been pulled anteriorly and caudally and then secured by tape to the arm board. Looking at the head, it is evident that the vertex is tilted down to raise the mastoid process as much as possible. (**c**) Patient in Fukushima position. Note the roll (red arrowhead) positioned in the axilla to prevent any compression of the brachial plexus. The Mayfield device (blue arrowhead) has been positioned, and the frontal pin is set inside the hairline. Note that the head is elevated and flexed (curved grey arrow) so that the chin is at a distance of two-finger width from the sternum. (**d**) Patient in Fukushima position (particular). Note that the head is rotated contralaterally (curved grey arrow) to the lesion, the vertex is tilted down (curved orange arrow), and, as a result, the mastoid tip is set as the top of the surgical field (red arrowhead). *Figure (a) reprinted from T. Fukushima, A. Friedman, L. Mastronardi, T. Sameshima, Fukushima's Microanatomy and Dissection of the Temporal Bone – Second Edition, 2007, with permission from AF-Neurovideo, Inc.*

avoid compression of the sciatic nerve. Soft padding is positioned on the chest wall and in the both axillas to avoid compression of the brachial plexus.

Both arms are outstretched on padded arm boards to prevent the compression of the ulnar nerve at the medial epicondyle of the nondependent arm and of the radial nerve at the radial groove of the humerus of the dependent arm. The dependent arm is positioned at 90° to the longitudinal axis of the patient's torso—although it may also be flexed at the elbow—while the upper arm is positioned at 45° to the longitudinal axis. In positioning the nondependent arm, the shoulder is brought anteriorly and gently pulled in the caudal direction (Fig. 3.1b). Then, it is secured by tape to the arm board. This maneuver pulls the shoulder away from the surgeon and allows a broader exposure.

The head is placed in three-point fixation head holder (Mayfield device) so that the two posterior pins are at the inion and at the mastoid body, respectively, while the frontal one is positioned inside the hairline for a better cosmetic result (Fig. 3.1c). The head is rotated contralaterally to the lesion so that the petrous ridge is perpendicular to the floor and the internal acoustic meatus (IAM) is in line of sight. The head is flexed so that the chin is at a distance of two-finger width from the sternum—this maneuver brings the mastoid process further away from the ipsilateral shoulder. Then, the head is elevated to keep one-hand space from the table, and the vertex is tilted down so that the mastoid process is at the highest point of the operative field (Fig. 3.1d).

3.2   Supine Position [1–3]

The supine position is one of the most commonly used and versatile positions in neurosurgery as it requires little manipulation of the patient and the tubes.

Following intubation, while the patient lies in supine position, the head attachment of the operating table is removed, and the head is placed in three-point fixation head holder. Dual pins are placed in coronal plane superior and inferior to the contralateral ear, and the single pin is positioned on the side of the lesion. The head fixation device is secured to the operating table using the table attachment arm. At this point, the head is tilted at 90° contralaterally to the lesion so as to expose the surgical site. Extreme head rotation poses a major risk of jugular obstruction and tracheal tube obstruction or displacement. Particular care must be taken while performing this maneuver, and the anesthesiologist's surveillance is always recommended. To relieve the neck rotation, a small pillow may be placed between the scapulae to allow the neck to extend slightly. In addition, the head may be flexed in order to set the mastoid process apart from the shoulder.

Once the head is fixed, the patient is secured to the operating table with foam padding and tape. A pillow is placed under the knees to prevent sciatic nerve stretching, and foam padding is positioned under the heels so as to avoid pressure ulcers. The arms are accordingly secured with foam padding and tape at the patient's sides. All the bony prominences are paddled with foam to avoid bedsores or nerve compression, as in the case of the ulnar grooves.

Finally, the operating table may be manipulated and put in reverse Trendelenburg position to allow better venous drainage and brain relaxation. Notably, the reverse Trendelenburg position sets the head above the heart and might increase the risk of venous air embolism (VAE).

3.3   Semi-sitting Position [1, 3, 4]

The semi-sitting position is a modern variant of the original sitting position, used at the beginning of the twentieth century and then progressively fallen into disuse due to concerns about its main medical complication, which is the VAE. On the surgical side, the semi-sitting position allows optimal exposure as the intracranial pressure (ICP) decreases and the gravity ensures better drainage of blood and CSF away from the surgical field [5–11].

Following intubation, the patient's head is fixed in three-point fixation head holder: the dual pins are positioned in axial plane superior and anterior to the ear on the non-affected side, and the single pin is placed on the affected side, around the linea temporalis and anterior to the external acoustic meatus.

The operating table is then manipulated. The back is elevated and flexed so that the patient reaches a sitting position with the hips flexed. To avoid sudden hypotension or hemodynamic instability, this maneuver should be performed slowly and under the anesthesiologist's supervision. The patient's knees should be flexed, too, by placing a pillow underneath so as to avoid sciatic nerve stretching. The caudal end of the operating table is elevated in order to increase the venous return to the heart.

The head fixation device is secured to the table with an appropriate adaptor. The patient's head is flexed to ensure optimal visualization of the surgical site. Ideally, this maneuver would set the tentorium parallel to the floor. Excessive head flexion has been addressed as the main cause for the reported cases of lingual, palatal, and laryngeal edema after the use of this position as it facilitates the displacement of and the compression by the tracheal tube.

The patient is secured to the operating table with padded safety belt or foam padding and tape. It is important to ensure that the patient's body is well supported and does not hang from the head in the fixation device, which would cause neck traction and pose a threat to the integrity of the spinal cord. The few reported cases of quadriplegia following the use of the semi-sitting position may be explained by this mechanism. The arms are either secured in a neutral position, padded and flexed across the abdomen, or placed on arm boards with slight flexion at the elbows. The bony prominences are obviously padded to avoid the risk of pressure ulcers.

3.4 Comparison of Outcomes Following Operations Performed in the Three Positions

The main advantages and disadvantages of the semi-sitting position are due to the effects of gravity. In fact, as a venous gradient develops between the head and the heart, the drainage of both blood and CSF improves and results in almost no blood during craniotomy and dissection. As a consequence, no continuous suction is needed, and the surgeon can use a bimanual dissection technique to approach the lesion. Finally, gravity itself retracts the cerebellar hemispheres so that no retraction is needed—although with the risk of prolapse of the cerebellum. However, the pros of the semi-sitting position have to be contrasted with the fact that a venous gradient facilitates air aspiration into blood vessels, thus causing VAE [7]. This potentially catastrophical intraoperative complication is also pathophysiologically linked to two other life-threatening conditions: tension pneumocephalus and paradoxical air embolism (PAE). Tension pneumocephalus is induced by the aspiration of air into epidural or intradural spaces in sufficient volumes to exert a mass effect, thus raising ICP and the risk of herniation. PAE is due to the obstruction of an arterial vessel by air passing through a right-to-left shunt, namely, a patent foramen ovale (PFO) [10].

The controversy over the balance between benefits and risks in the use of semi-sitting position has led to the growth of extensive literature about intra- and postoperative outcomes comparison between different positions in general neurosurgery and in VS surgery.

Rath et al. examined the complications related to positioning in posterior fossa surgery [11] and found that the incidence of VAE was significantly higher in the semi-sitting position (15.2%) than in the supine position (1.4%), which also carried the risk for higher blood loss. An advantage reported about the semi-sitting position was a significantly higher rate of preservation of lower cranial nerves, although biased by the diverse range of different operations taken into consideration in the study. No significant differences were found in postoperative complication rates, which led the authors to conclude that both the positions would be safe as long as good patient screening and preparation is performed before using the semi-sitting position. Fathi et al. confirmed this thesis [12] in their systematic review about air embolism complications of neurosurgery in semi-sitting position. In addition, they underlined the importance of screening for PFO detection as its prevalence in neurosurgical patients is not negligible (5–33%). Thus, numerous studies have been conducted to propose a correct algorithm of patient selection, and in spite of the lack of official guidelines, specific measures have been addressed as appropriate and necessary [5, 8, 13]: preoperative radiographs of cervical spine to exclude neck instability, preoperative trans-esophageal echocardiography (TOE) to detect PFO, and intraoperative monitoring to prevent, diagnose, and promptly manage VAE. Whether PFO is an absolute or relative contraindication to the use of semi-sitting position is still controversial, and there is no consensus over the gold standard technique of intraoperative monitoring—although TOE has shown the highest sensitivity in detecting even not clinically significant episodes of VAE (25.6% in Ganslandt et al. [9]).

As far as VS surgery is concerned, the semi-sitting position has been compared to the supine and the lateral positions not only in terms of safety but also in terms of specific outcomes such as extent of resection, facial nerve (N VII) preservation, and hearing preservation (HP). However, the results are inconclusive and generally biased by both the experience of the surgeons conducting the studies and the policies of the respective institutions. Spektor et al. compared semi-sitting position and lateral position and reported that N VII preservation was correlated to the extent of resection rather than to the preferred position [14]. The correlation between extent of resection and position was reported as not significant. In addition, the only significant differences found were related to the preparation and duration of the operations (shorter when lateral position was preferred). On the contrary, Roessler et al. reported opposite results in their study [15]: they found that operations performed using the semi-sitting position were shorter and more effective in comparison with the lateral position. In fact, at 6 months of follow-up, N VII preservation rate was higher in the former group (63% vs 40%) as well as HP rate (44% vs 14%). However, the authors addressed the retrospective nature of their study and the individual skills of the different surgeons as confounding factors.

Although no randomized trials are available in the literature to answer the question about which is the best position to choose in VS surgery, it can be surely assumed that the three positions are safe [6, 16, 17] but may not be appropriate in all the cases. The use of semi-sitting position requires highly specialized neuroanesthesiologists and continuous intraoperative TOE which is expensive and not necessarily available in all the departments. As illustrated by the opposite results of the studies by Spektor et al. and Roessler et al., the surgeon's experience and preferences may strongly impact on the intra- and postoperative outcomes, even more than the intrinsic characteristics of the position itself. As a consequence, patient positioning should be based on surgical team preference.

Despite of the experience of outstanding authors using the semi-sitting position, in our daily practice, we routinely use the lateral Fukushima position: the mean duration of surgery is about 4 h and is mainly in relation to the size of tumor and adherence of capsule to facial and cochlear nerves and to brainstem. As reported in other chapters, the results are in line with those of the current international literature.

References

1. Di Ieva A, Lee JM, Cusimano MD. Handbook of skull base surgery. New York: Thieme; 2016. xxvii, 978 p.
2. Sameshima T. Fukushima's microanatomy and dissection of the temporal bone. 2nd ed. In: Sameshima T, editor. Raleigh: AF-Neurovideo, Inc.; 2007. 115 p.
3. Winn HR. Youmans and Winn neurological surgery. 7th ed. Philadelphia: Elsevier; 2017.
4. Quiñones-Hinojosa A, Rincon-Torroella J. Video atlas of neurosurgery: contemporary tumor and skull base surgery. 1st ed. Edinburgh and New York: Elsevier; 2017. xxx, 285 p.
5. Ammirati M, Lamki TT, Shaw AB, Forde B, Nakano I, Mani M. A streamlined protocol for the use of the semi-sitting position in neurosurgery: a report on 48 consecutive procedures. J Clin Neurosci. 2013;20(1):32–4.
6. Boublata L, Belahreche M, Ouchtati R, Shabhay Z, Boutiah L, Kabache M, et al. Facial nerve function and quality of resection in large and giant vestibular schwannomas surgery operated by retrosigmoid transmeatal approach in semi-sitting position with intraoperative facial nerve monitoring. World Neurosurg. 2017;103:231–40.
7. Duke DA, Lynch JJ, Harner SG, Faust RJ, Ebersold MJ. Venous air embolism in sitting and supine patients undergoing vestibular schwannoma resection. Neurosurgery. 1998;42(6):1282–6; discussion 6–7.
8. Gale T, Leslie K. Anaesthesia for neurosurgery in the sitting position. J Clin Neurosci. 2004;11(7):693–6.
9. Ganslandt O, Merkel A, Schmitt H, Tzabazis A, Buchfelder M, Eyupoglu I, et al. The sitting position in neurosurgery: indications, complications and results. A single institution experience of 600 cases. Acta Neurochir. 2013;155(10):1887–93.
10. Porter JM, Pidgeon C, Cunningham AJ. The sitting position in neurosurgery: a critical appraisal. Br J Anaesth. 1999;82(1):117–28.
11. Rath GP, Bithal PK, Chaturvedi A, Dash HH. Complications related to positioning in posterior fossa craniectomy. J Clin Neurosci. 2007;14(6):520–5.
12. Fathi AR, Eshtehardi P, Meier B. Patent foramen ovale and neurosurgery in sitting position: a systematic review. Br J Anaesth. 2009;102(5):588–96.

13. Günther F, Frank P, Nakamura M, Hermann EJ, Palmaers T. Venous air embolism in the sitting position in cranial neurosurgery: incidence and severity according to the used monitoring. Acta Neurochir. 2017;159(2):339–46.
14. Spektor S, Fraifeld S, Margolin E, Saseedharan S, Eimerl D, Umansky F. Comparison of outcomes following complex posterior fossa surgery performed in the sitting versus lateral position. J Clin Neurosci. 2015;22(4):705–12.
15. Roessler K, Krawagna M, Bischoff B, Rampp S, Ganslandt O, Iro H, et al. Improved postoperative facial nerve and hearing function in retrosigmoid vestibular schwannoma surgery significantly associated with semisitting position. World Neurosurg. 2016;87:290–7.
16. Cardoso AC, Fernandes YB, Ramina R, Borges G. Acoustic neuroma (vestibular schwannoma): surgical results on 240 patients operated on dorsal decubitus position. Arq Neuropsiquiatr. 2007;65(3A):605–9.
17. Kaye AH, Leslie K. The sitting position for neurosurgery: yet another case series confirming safety. World Neurosurg. 2012;77(1):42–3.

Instrumentation for Acoustic Neuroma Microneurosurgery

4

Luciano Mastronardi, Alberto Campione,
Guglielmo Cacciotti, Raffaelino Roperto, Fabio Crescenzi,
Ali Zomorodi, and Takanori Fukushima

Vestibular schwannoma (VS) surgery belongs to the wide field of skull base surgery, which is recognized as technically demanding due to extensive manipulation of anatomical structures within a narrow space. The surgical instrumentation employed in such procedures has developed in the last decades and benefited from contributions by both authoritative surgeons—as in the case of the senior author of this book—and pioneering medical device companies.

The most important instruments for VS surgery can be grouped as follows:

- Standard microsurgical instruments
- Electronic microsurgical devices
- Endoscopes
- Intraoperative neurophysiological monitoring (IONM) devices

4.1 Standard Microsurgical Instruments

"Mechanical" microsurgical instruments are defined as "standard" as opposed to devices that are used as well for tumor manipulation and are electronically activated.

The main steps of VS manipulation are dissection, fragmentation, and piecemeal resection.

L. Mastronardi (✉) · A. Campione · G. Cacciotti · R. Roperto · F. Crescenzi
Department of Neurosurgery, San Filippo Neri Hospital—ASLRoma1, Rome, Italy
e-mail: mastro@tin.it

A. Zomorodi · T. Fukushima
Division of Neurosurgery, Duke University Medical Center, Carolina Neuroscience Institute, Raleigh, NC, USA
e-mail: ali.zomorodi@duke.edu; Fukushima@carolinaneuroscience.com

© Springer Nature Switzerland AG 2019
L. Mastronardi et al. (eds.), *Advances in Vestibular Schwannoma Microneurosurgery*, https://doi.org/10.1007/978-3-030-03167-1_4

Dissection consists of tumor detachment from surrounding dura, arachnoid, or neurovascular structures. In cases of large VS, a 2-mm tip, 14-mm meso-type brain spatula may be used to gently hold the brain to facilitate microdissection in the depths. In order to expose a cleavage plane, the tumor capsule or the arachnoid should be gripped with forceps so as to stretch arachnoid strands which are to be cut with micro scissors. Thus, *1-mm micro alligator tumor forceps* and *thin blade micro scissors (straight, curved, Kamiyama type)* are required for capsule dissection [1, 2] (Fig. 4.1).

Once the internal auditory canal (IAC) has been exposed, the dissection of the canalicular dura mater and the cranial nerves inside requires nine pieces of supermicro dissectors that are essential to eradicate the tumor: 90°, 70°, and 45° Hitzelberger-McElveen knives, 90° and 45° sharp hook knives, micro sickle knife, sharp dog dissector, and 90° micro cup curette of 0.75 mm and 1 mm. The *sharp dog dissector* allows for gentle separation of neurovascular structures without damage [1, 2] (Fig. 4.2).

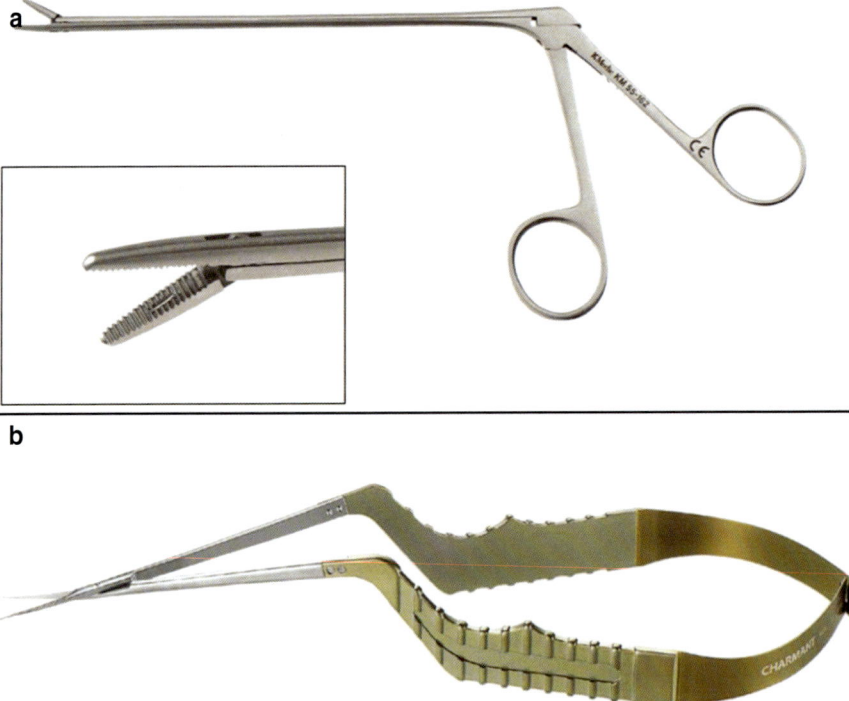

Fig. 4.1 (**a**) Micro alligator tumor forceps; in the bottom left box, the tip is enlarged. (**b**) Thin blade micro scissors, bayonet blade (*courtesy of Charmant Inc. | Medical division, 6-1 Kawasari-cho, Sabae City, Fukui-pref 916-8555, Japan*)

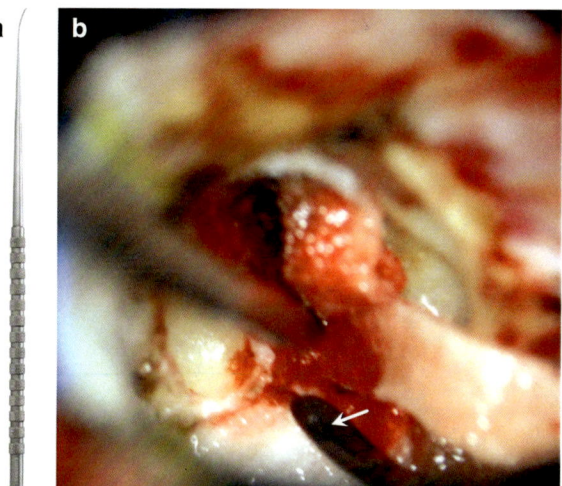

Fig. 4.2 Sharp dog dissector. (**a**) Sharp dog dissector, scale 0.66:1. (**b**) Dissection of facial nerve and cochlear nerve at the internal acoustic meatus. The white arrow indicates the tip of the sharp dog dissector (*courtesy of Charmant Inc. | Medical division, 6-1 Kawasari-cho, Sabae City, Fukui-pref 916-8555, Japan*)

Microsurgical knives are required for dural incisions, tumor fragmentation, and debulking; the most important ones are the *Hitzelberger-McElveen knife (bullet tip)*, the *45° and 90° sharp hook knives*, and the *sickle knife* [1, 2] (Fig. 4.3).

The piecemeal resection of the tumor, especially in a narrow space such as the IAC, requires microsurgical curettes that are used to collect tumor fragments. The most important types are *ring curettes* of different dimensions and *cup curettes* [1, 2] (Fig. 4.4).

4.2   Electronic Microsurgical Devices

The most important element of VS surgery is precise hemostasis to maintain the bloodless, clean, dry operative field. In order to achieve precise hemostasis, the surgeon must have three *bipolar forceps*: one is keyhole SILVERGlide bipolar 0.4 mm tip, which provides the most efficient coagulation of the highly vascularized VS. Also needed is a Tokyo-designed microbipolar with 2 mm tip exposure, 0.2–0.3 mm tip and straight type, 15°, 30°, and 45° up-angled microbipolar system. Lastly, micro-bipolar forceps of Italian design is needed, which provides the maximum keyhole style with 0.15, 0.2, 0.25, and 0.3 micro tips made of silver alloy.

A *handheld 2µ-thulium flexible laser fiber* is an innovative surgical instrument that can be used for hemostasis, capsule and dural incision and vaporization, and tumor debulking. The device we use in our unit is RevoLix™, Lisa laser products, Katlenburg-Lindau, Germany [3, 4] (Fig. 4.5). 2µ-thulium laser has a wavelength of 2 micron; this allows for excess laser radiation to be absorbed by the irrigation so

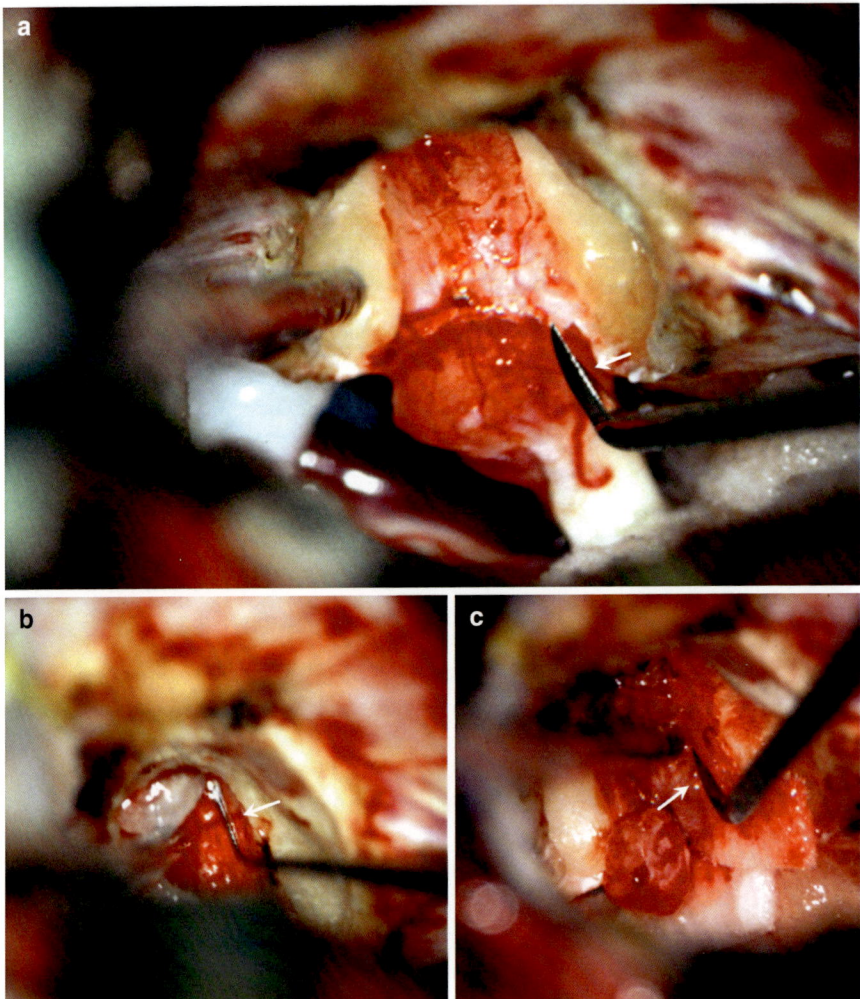

Fig. 4.3 Microsurgical knives. (**a**) The IAC is cut after bone drilling. The white arrow indicates the tip of a 45° sharp hook knife. (**b**) Dissection of facial nerve and cochlear nerve at the internal acoustic meatus. The white arrow indicates the tip of a Hitzelberger-McElveen knife (bullet tip). (**c**) Dissection of the IAC. The white arrow indicates a 90° sharp hook knife

that it does not affect tissue more than 3 mm from the tip of the fiber. Tissue damage is limited to 0.2–1.0 mm around, and the minimal width of the fiber allows for perfect visualization and control of the surgical field. In addition, continuous light emission avoids the explosive effects of pulsed wave lasers.

The *ultrasonic aspirator* is mainly used for tumor debulking and opening of the IAC. The device we use in our unit is Sonopet®, Stryker, Kalamazoo, MI, USA [3, 4] (Fig. 4.6). The physical principle at the base of its functioning is that the vibration

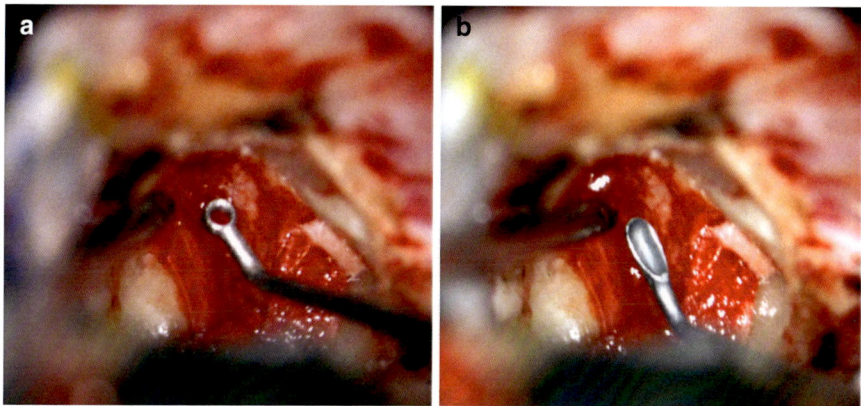

Fig. 4.4 Microsurgical curettes. (**a**) Tip of a 1-mm ring curette. (**b**) Tip of a cup curette

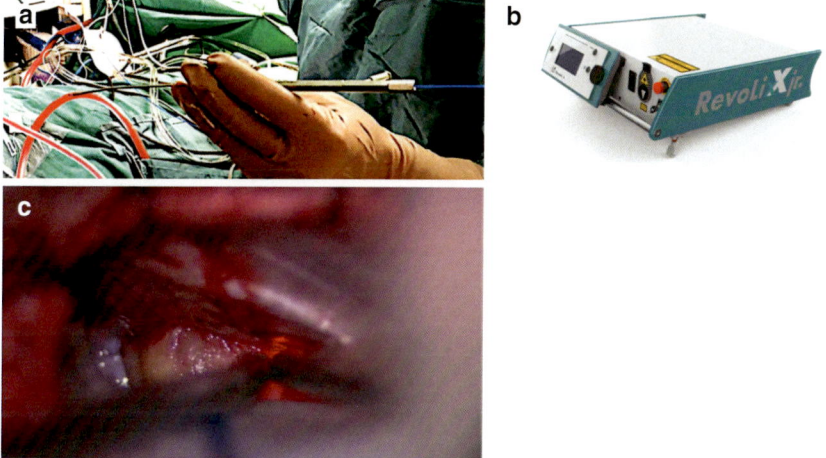

Fig. 4.5 Handheld 2μ-thulium flexible laser fiber. (**a**) Flexible silica fiber probe. (**b**) Console, RevoLix jr. (*courtesy of LISA laser products OHG, Fuhrberg & Teichmann, Albert-Einstein-Str. 1-9D-37191 Katlenburg-Lindau, Germany*). (**c**) Laser employed during surgery

that it provides consists of alternate high- and low-pressure peaks delivered on targeted tissue. Cells expand under negative pressures and peaks of high pressure cause them to burst. The process is selective because tissues with high water content are more susceptible to cavitation. Collagen and elastin fibers vibrate in resonance with the acoustic vibrations so that blood vessels and nerves are eventually largely untouched. Dedicated cutting ablation tips are required to overcome the phenomenon of resonance in calcified and fibrotic tissues: the honed edges of such tips break the collagen bonds in stiff tissues and allow for cavitation.

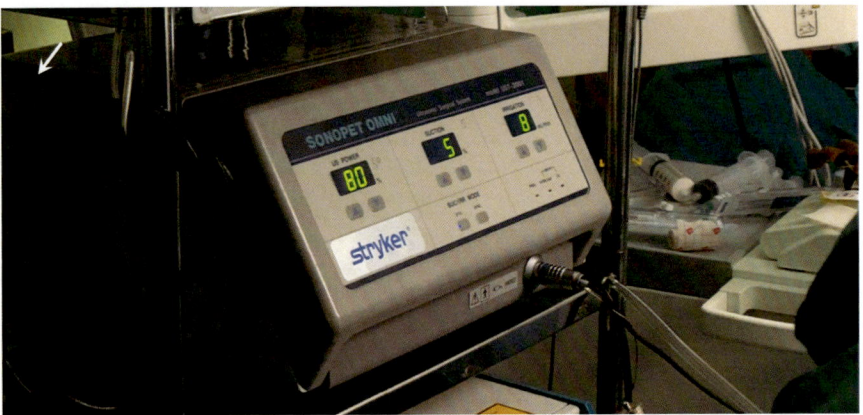

Fig. 4.6 Ultrasonic aspirator. Console showing values of power, suction, and irrigation (Sonopet®, Stryker, Kalamazoo, MI, USA)

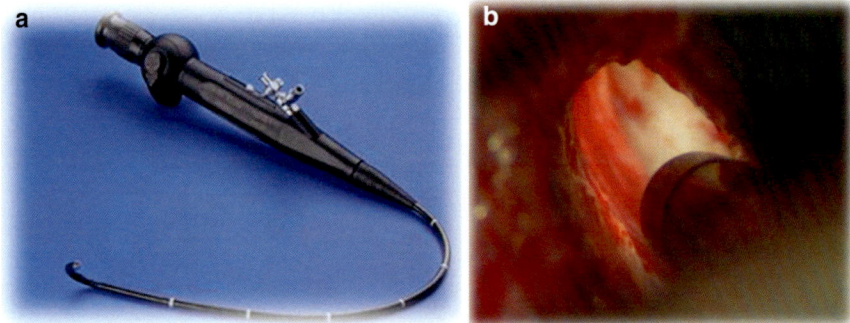

Fig. 4.7 (**a**) Flexible endoscope. (**b**) Flexible endoscope introduced into the IAC (*reprinted from World Neurosurgery, 115, Francesco Corrivetti, Guglielmo Cacciotti, Carlo Giacobbo Scavo, Raffaelino Roperto, Luciano Mastronardi, Flexible Endoscopic-Assisted Microsurgical Radical Resection of Intracanalicular Vestibular Schwannomas by a Retrosigmoid Approach: Operative Technique, Pages No. 229.233, 2018, with permission from Elsevier*)

4.3   Endoscopes

Endoscope-assisted microsurgery allows complete VS resection as the surgeon is enabled to directly observe the lateral extremity of the IAC, which is a "zone around the corner" for the straight view provided by the surgical microscope. *30°–70° rigid endoscopes* have been widely used by diverse teams [5]; however, we have recently begun using a 4-mm flexible video endoscope (4 mm × 65 cm, Karl Storz, Inc.) (Fig. 4.7). The main advantage of *flexible endoscopes* is that the tip can be electronically deflected by a proximal joystick control: this allows for an easy surgical exploration and a face-to-face visualization of tumor remnants.

4.4 Intraoperative Neurophysiological Monitoring (IONM) Devices

Facial nerve intraoperative monitoring consists of both direct electrical stimulation and free-running electromyography (EMG) recording. The electrical stimulus is provided by a central console and delivered at the target nerve through either a monopolar or a bipolar probe; the response is then recorded and shown on the console or on a screen connected to it. The device we use in our center is *Nimbus I-Care, Innopsys, Carbonne, France* [3, 4, 6, 7]. The senior author uses a different device in his institution, *NIM-Neuro® 3.0, Medtronic, Minneapolis, MN, USA*. A general principle, as regards the use of probes, is to employ the monopolar probe for gross stimulation to vaguely locate the facial nerve when it has not been exposed yet; instead, the bipolar probe should be used to verify the anatomical and functional preservation of the nerve.

Intraoperative monitoring of the cochlear nerve is usually performed by evoking and recording acoustic brainstem responses (ABR), which are elicited by specific stimuli such as tone bursts and clicks. However, these sounds are composite by nature, and as the cochlea is tonotopically organized, the basilar membrane is stimulated at different sites in different moments; this results in ABR temporal smearing and difficult neurophysiological monitoring. The CE-Chirp® stimulus family emerged as an innovative factor in the field: in this case, the stimulus is designed to stimulate all the desired basal membrane regions at the same time, which results in time sparing and waves of larger amplitude, i.e., easier to analyze [6, 7]. The device we use in our center is *Eclipse EP15 ABR system, Interacoustics, Middelfart, Denmark*, and it employs CE-Chirp® stimulation protocol.

References

1. Sameshima T, Mastronardi L, Friedman AH, Fukushima T. Microanatomy and dissection of temporal bone for surgery of acoustic neuroma and petroclival meningioma. 2nd ed. Raleigh: AF Neurovideo, Inc.; 2007.
2. Wanibuchi M, Fukushima T, Friedman AH, Watanabe K, Akiyama Y, Mikami T, et al. Hearing preservation surgery for vestibular schwannomas via the retrosigmoid transmeatal approach: surgical tips. Neurosurg Rev. 2014;37(3):431–44; discussion 44.
3. Mastronardi L, Cacciotti G, Roperto R, Tonelli MP, Carpineta E. How I do it: the role of flexible hand-held 2μ-thulium laser Fiber in microsurgical removal of acoustic neuromas. J Neurol Surg B Skull Base. 2017;78(4):301–7.
4. Mastronardi L, Cacciotti G, Scipio ED, Parziale G, Roperto R, Tonelli MP, et al. Safety and usefulness of flexible hand-held laser fibers in microsurgical removal of acoustic neuromas (vestibular schwannomas). Clin Neurol Neurosurg. 2016;145:35–40.
5. Tatagiba MS, Roser F, Hirt B, Ebner FH. The retrosigmoid endoscopic approach for cerebellopontine-angle tumors and microvascular decompression. World Neurosurg. 2014;82(6 Suppl):S171–6.
6. Di Scipio E, Mastronardi L. CE-Chirp® ABR in cerebellopontine angle surgery neuromonitoring: technical assessment in four cases. Neurosurg Rev. 2015;38(2):381–4; discussion 4.
7. Mastronardi L, Di Scipio E, Cacciotti G, Roperto R. Vestibular schwannoma and hearing preservation: usefulness of level specific CE-Chirp ABR monitoring. A retrospective study on 25 cases with preoperative socially useful hearing. Clin Neurol Neurosurg. 2018;165:108–15.

Retrosigmoid Approach

5

Luciano Mastronardi, Alberto Campione,
Guglielmo Cacciotti, Raffaelino Roperto,
Carlo Giacobbo Scavo, Ali Zomorodi,
and Takanori Fukushima

The retrosigmoid (RS) approach provides direct access to and good control of neurovascular structures in the cerebellopontine angle (CPA); it is the most commonly performed approach when attempting hearing preservation (HP) in vestibular schwannoma (VS) surgery, regardless of the tumor size. However, it is still a matter of debate whether RS or middle fossa approach (MF, not discussed here) should be preferred in case of good preoperative hearing.

In 2006, Samii et al. [1] reported total removal in 98% of 200 cases, good-to-excellent long-term N VII function in 81%, and HP in 51%. They concluded that total microsurgical removal of small VS (<20 mm) by RS approach is feasible and curative in one stage, with good preservation of neurological functions, including hearing in patients with preoperative socially useful hearing (SUH). In 592 patients, Wanibuchi et al. [2] reported HP in 53.7% of large VS (diameter >20 mm) and 74.1% of all sizes. Scheller et al. [3] studied long-term stability of HP and regeneration capacity of cochlear nerve in 112 VSs operated on by RS approach; in particular, they investigated efficacy of prophylactic parenteral nimodipine, without clinically relevant effects on preservation of cochlear function. They did not find any significant change in HP between early and 1-year control, concluding that result of early postoperative hearing performance is a reliable prognostic factor for future hearing ability.

Peng and Wilkinson [4] maintained that in patients younger than 65 with small VSs, microsurgery by middle fossa (MF) approach ensures long-term HP. Satar et al. [5] reviewed 11 studies reporting effects of tumor size on hearing (1073 cases) and N VII function (797 cases) after MF approach. Their meta-analysis showed that tumor

L. Mastronardi (✉) · A. Campione · G. Cacciotti · R. Roperto · C. Giacobbo Scavo
Department of Neurosurgery, San Filippo Neri Hospital—ASLRoma1, Rome, Italy
e-mail: mastro@tin.it

A. Zomorodi · T. Fukushima
Division of Neurosurgery, Duke University Medical Center, Carolina Neuroscience Institute, Raleigh, NC, USA
e-mail: ali.zomorodi@duke.edu; Fukushima@carolinaneuroscience.com

© Springer Nature Switzerland AG 2019
L. Mastronardi et al. (eds.), *Advances in Vestibular Schwannoma Microneurosurgery*, https://doi.org/10.1007/978-3-030-03167-1_5

size (including intracanalicular portion) is the main predictor of hearing and N VII outcome. On 78 VSs with maximum diameter ≤2 cm, operated on by RS, translaby-rinthine, or MF approaches, Anaizi et al. [6] reported 95% House Brackmann Grade I (HBI) or HBII N VII function at a mean follow-up of about 3 years and 36% service-able hearing. Sameshima et al. [7] compared RS and MF approaches for HP in 504 VSs <1.5 cm: SUH preservation was 76.7% in MF approach patients and 73.2% in RS (p = NS). Temporary N VII deficit was observed more frequently after MF approach (p < 0.03), with good recovery in both subgroups. In their series, about 14% of MF cases had transient symptoms of temporal lobe edema; no cerebellar side effects were reported in RS cases. They concluded that although 1-year hearing and N VII func-tions were similar, RS approach had advantages over MF. Therefore, although wait-and-scan and SRS have an established role, microsurgery by RS approach is a safe option for small VS, with low morbidity and good N VII and HP results.

5.1 Patient Preparation, Positioning, and Incision

In the contemporary most advanced VS surgery, the cardinal importance is the use of a continuous lumbar spinal CSF drainage tube, which is the most efficient technique of achieving a slack brain during surgery and postoperative wound closure (draining around 10 cm^3/h) [8–10]. Routinely, advanced neuro-anesthe-sia is used with hyperventilation, dexamethasone 10–20 mg, and mannitol 50–100 mg. The intraoperative monitoring devices such as somatosensory poten-tials, facial nerve (N VII) EMG, and brainstem evoked auditory responses (BAERs)—when HP is attempted—are activated, and the electrodes are ade-quately positioned.

The patient is placed in Fukushima position (see Chap. 3) and superficial land-marks are located (Fig. 5.1). The root of zygoma is palpated and marked with a

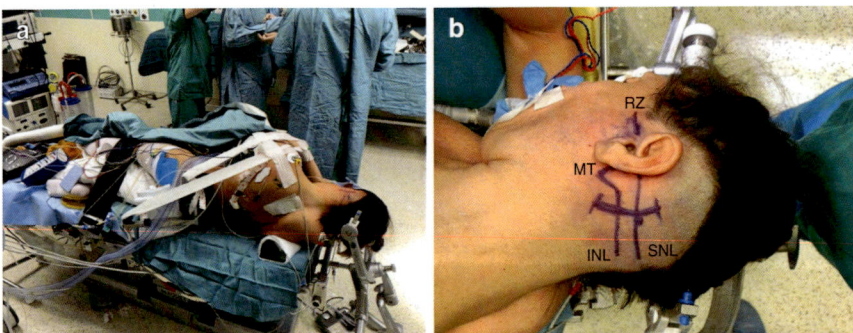

Fig. 5.1 Patient positioning and superficial landmarks location. (**a**) Patient placed in Fukushima position. (**b**) Superficial landmarks of the region. The curved posterior to the pinna is the incision line, and its intersection with SNL is the superficial projection of the asterion and (deep) the genu between transverse sinus and sigmoid sinus. For a detailed description, see the text. *RZ* root of zygoma, *MT* mastoid tip, *SNL* superior nuchal line, *INL* inferior nuchal line

Fig. 5.2 Free pericranial flap harvesting for final dural closure

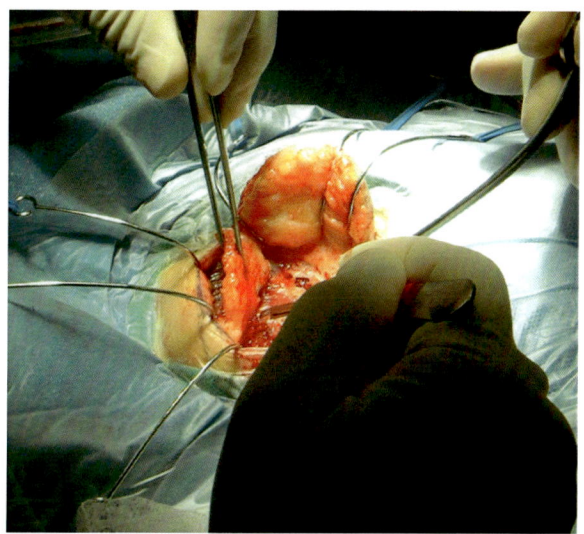

horizontal line whose posterior prosecution is the superficial projection of the superior nuchal line (SNL). The mastoid tip is palpated and marked as well; the horizontal line that runs posteriorly to the mastoid body is the superficial projection of the inferior nuchal line (INL). SNL and INL serve, respectively, as superior and inferior margins of the projection of the transverse sinus. A retroauricular "C"-shaped or slight-curved incision measuring about 5 cm is performed; it runs from the supra-mastoid crest to the level of the mastoid tip and passes 2 cm posterior to the mastoid body [2, 11, 12]. The intersection between the incision line and the SNL locates the asterion and the genu between transverse sinus and sigmoid sinus, which constitute the superior and anterior limits of the craniotomy, respectively.

After the incision, the scalp is elevated in two layers in order to maintain the musculofascial-periosteal layer [2, 11, 12]. A free pericranial flap (about 3 cm × 3 cm) for dural closure is harvested (Fig. 5.2); during the remainder of the procedure, the graft patch is soaked and stored in gentamycin-enriched saline solution [13]. The posterior neck muscles are incised along the line of the skin incision and elevated from the suboccipital surface. The skin flap and musculofascial layers are retracted anteriorly with blunt-tip hooks. Then, the bone is meticulously exposed and the sutures are identified.

5.2   Keyhole Retrosigmoid Craniotomy

A linear drilling is made using a 4 or 5-mm extra-coarse diamond drill starting at the inferior corner of the digastric groove (see (1) in Fig. 5.3a). After exposing the intact dura, a 5-mm longitudinal groove is made at the posterior border of the mastoid body, exposing safely the sigmoid sinus (see (2) in Fig. 5.3a). Then, the groove is continued downward along the inferior edge of the proposed bone flap (suboccipital

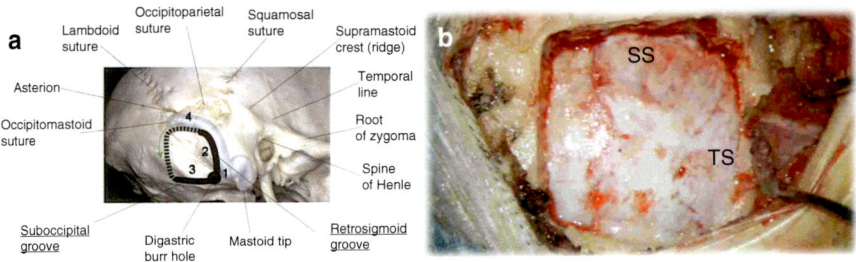

Fig. 5.3 Retrosigmoid craniotomy (right side). (**a**) The margins of the craniotomy are highlighted, and the steps to isolate the bone flap are marked by numbers (see the text for description). (**b**) The bone flap has been removed; the sigmoid sinus (SS) represents the cranial edge of the craniotomy, while the transverse sinus (TS) is the anterior limit (*Fig. (a) reprinted from T. Fukushima, A. Friedman, L. Mastronardi, T. Sameshima, Fukushima's Microanatomy and Dissection of the Temporal Bone – Second Edition, 2007, with permission from AF-Neurovideo, Inc.*)

groove) (see (3) in Fig. 5.3a), as well as along the superior margin to expose the junction of the sigmoid sinus and the transverse sinus (see (4) in Fig. 5.3a), using a 4-mm extra-coarse diamond drill [11]. The entire dura is elevated around the bone margin, and the bone flap is therefore removed (Fig. 5.3b), resulting in a 3 cm × 3 cm craniotomy.

5.3 Dural Opening and Lateral Medullary Cistern CSF Aspiration

The dura is opened in a semicircular shape, covering part of the cerebellar hemisphere in order to serve as a protective sheath during retraction [2, 8, 9] (Fig. 5.4). The arachnoid membrane of the lateral medullary cistern is opened, and the CSF is aspirated for cerebellar relaxation [1, 2, 8, 9, 11, 12, 14]. A 2-mm tip, brain spatula is then inserted and fixed by a retractor so as to hold (not retract) the cerebellum and gain insight into the CPA.

5.4 Opening of the Internal Auditory Canal

The porus acusticus should be identified on the posterior surface of the pyramid in order to proceed with the opening of the internal auditory canal (IAC); however, in case of large VSs, it can be difficult to see. A dural landmark is of particular aid. Several vertical foldings span between the jugular foramen (inferiorly) and extend 5–7 mm cranially: the linear level where all of the foldings end and the dura tightly adheres to the posterior wall of the temporal bone is the Tübingen line, which is the projection of the inferior limit of the IAC. Dural elevation and drilling along this landmark allow unroofing of IAC when the internal acoustic meatus is not visible [15].

Fig. 5.4 Dural opening. The black dashed line marks the classical dural incision; the red dashed line marks the reversal incision that allows dura to be used as a protective sheath covering the cerebellar hemisphere underneath during retraction (*Reprinted and modified from T. Fukushima, A. Friedman, L. Mastronardi, T. Sameshima, Fukushima's Microanatomy and Dissection of the Temporal Bone – Second Edition, 2007, with permission from AF-Neurovideo, Inc.*)

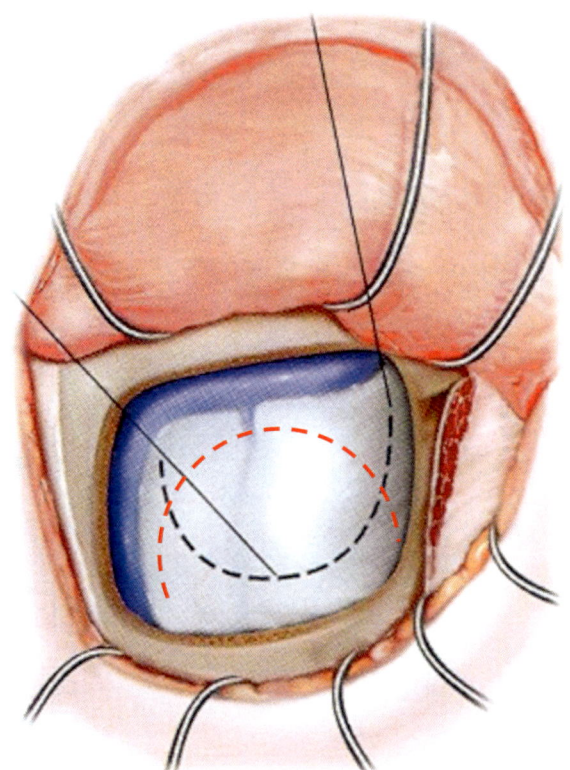

The dura posterior to the porus acusticus is either incised or removed by laser [8, 9] (Fig. 5.6a) (handheld 2μ-thulium flexible laser fiber, Revolix jr.®, Lisa laser USA, Pleasanton, CA, USA) in an inverted "U" shape, with base approximately at the fovea, which corresponds to the apex of the endolymphatic sac [2, 11]. The inverted U-shaped dural incision extends about 6–8 mm toward the fovea, and its base extends 2 mm above and 2 mm below the IAC [11] (Fig. 5.5). A problem in dural dissection may arise if anterior inferior cerebellar artery (AICA) is found firmly adherent to the petrous dura mater instead of running loosely within the CPA; a similar occurrence may be observed in 6% of patients. In these cases, both the dura and the AICA can be elevated conjointly and displaced medially: this procedure allows access to IAC without risk to the artery [16].

The canal is exposed with Sonopet Ultrasonic Aspirator (Stryker, Kalamazoo, MI) or drill; in this case, a 4-mm coarse diamond burr is used to expose the IAC, and progressively smaller diamond burrs are then used to define the lateral margins of the canal [2, 8, 9, 11]. It is important to remove the bone from the upper and lower corners of the porus acusticus to expose the facial and cochlear nerves, which are usually compressed by the tumor as it bulges out of the IAC [11]. Ebner et al. [17] found that, during the IAC exposure, the structure at major risk is the endolymphatic duct (ED) rather than the more posterolaterally located endolymphatic sac. The ED

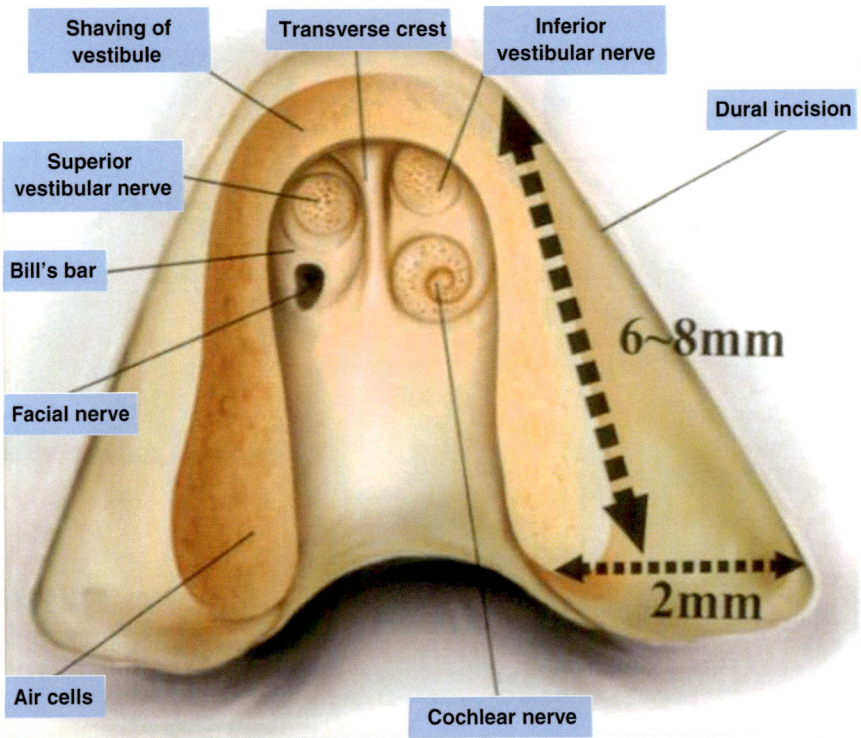

Fig. 5.5 Unroofed IAC on the right side. The disposition of the cranial nerves inside the canal can be appreciated. The tranverse crest is a bony edge that divides the canal into a superior compartment, occupied by superior vestibular nerve and facial nerve, and an inferior compartment, where the inferior vestibular nerve and the cochlear nerve are located. The Bill's bar is a crest that interposes between the facial nerve and the superior vestibular nerve so that N VII can be reliably located anterosuperiorly within the IAC if the normal anatomy is not distorted by the tumor. The dimensions here reported reflect the proposed size of dural incision. As illustrated, the IAC should be drilled until the transverse crest can be individuated (*Reprinted from T. Fukushima, A. Friedman, L. Mastronardi, T. Sameshima, Fukushima's Microanatomy and Dissection of the Temporal Bone – Second Edition, 2007, with permission from AF-Neurovideo, Inc.*)

runs within the vestibular aqueduct and connects the endolymphatic sac and the utriculosaccular duct. The distortion of the petrous bone topography due to VS jeopardizes the preservation of the ED, and an accurate preoperative planning is required. An imaginary line extending from the medial border of the sigmoid sinus to the fundus of the IAC determines whether the inner ear structures are at risk because of their location in the drilling direction: if the vestibular aqueduct crosses the line, a medial extension of the classic craniectomy has to be considered to open the working angle on the posterior meatal wall. If the possibility of drilling the IAC is very limited, an endoscope-assisted surgical plan should be arranged from the beginning.

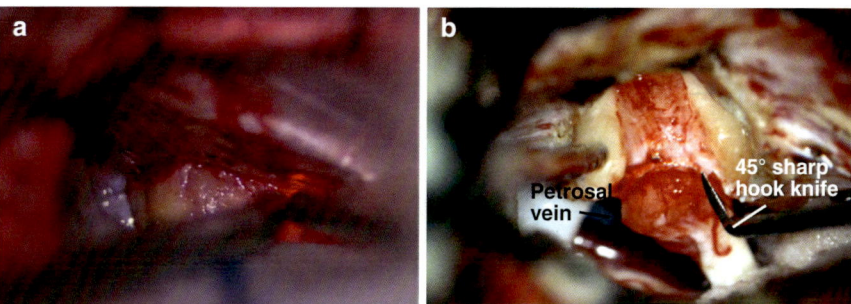

Fig. 5.6 (**a**) Dural removal by laser. (**b**) IAC dura being cut with a 45° sharp hook knife. Cranial to IAC is a petrosal vein that should always be spared during microsurgical dissection in order to avoid bleeding and to maintain the surgical field clean and clear

The IAC dura should be incised with a sharp knife along the edge of the bone so as not to leave an overhanging meningeal flap that may obstruct the surgeon's view [1, 2, 8, 9, 11, 14] (Fig. 5.6b). Thereafter, it is important to maintain the arachnoid plane around the tumor for successive steps of capsule dissection from surrounding neurovascular structures.

5.5   Localization of the Facial Nerve

The proximal N VII can be identified in the lateral aspect of the pontomedullary junction just under the choroid plexus of the lateral recess and the proximal portion of the glossopharyngeal nerve as the whitish band running over the surface of the lower pons (95% of cases). Then, the N VII will be shifting either straight upward, obliquely, or caudally along the ventral aspect of the tumor capsule. Then, the facial nerve can shift to the rostral dorsal level or ventral side. In the far ventral or anterior aspect of the large tumor, displacement, splitting, or fanning of the N VII is extremely variable, and the surgeon must be extremely careful with the final dissection of the tumor capsule from the thin N VII, particularly at the last 10 mm segment near the inferior margin of the enlarged IAC. The distal part of the N VII can be identified as the whitish thin band when the surgeon elevates the inferior and superior vestibular nerves and the tumor at the fundus of the IAC. The facial nerve can be identified and dissected as the whitish band while confirming its response by 0.1 mA of bipolar stimulus. The stimulation of N VII is performed with monopolar (on surface of tumor) or bipolar (close to the nerve) stimulator, starting from 2 mA or more (on the capsule for nerve course localization) to 0.3–0.05 mA (directly on the nerve for confirmation of function) [2, 8–10]. If the facial nerve locates 2 or 3 cm away in the ventral aspect of the tumor, a 5–10 mA stimulation may be needed to locate the response. If the capsule thickness becomes 1 or 1.5 cm, 2–3 mA may be suitable. If the capsule thickness is down to 1 cm, a stimulation of 0.07–1.2 mA is feasible. If the facial nerve response is 0.07 or 0.5 mA, tumor capsule becomes very thin, less than 5 mm of thickness. If the N VII responds to the 0.05–0.2 mA stimuli, the probe is already on the nerve. N VII is expected to be located anterior (31–52%), anterosuperior (38.5–48%), anteroinferior (5.3–21%), or, very rarely, dorsal (0.3–3.8%) to the tumor. N VII EMG allows for nerve anatomical and functional preservation (see Chap. 9: "Intraoperative identification and localization of facial nerve: position, course, and functional preservation").

5.6   Capsular Elevation and Tumor Debulking

According to the Fukushima's technique, a V-cut [2, 8–12] is usually performed on the dorsal surface of tumor with laser fiber [8, 9] or with microscissors and debulking of tumor obtained with microscissors, microcurettes, bipolar forceps, Sonopet Ultrasonic Aspirator, and handheld laser for vaporizing and cutting [8, 9] (Fig. 5.7).

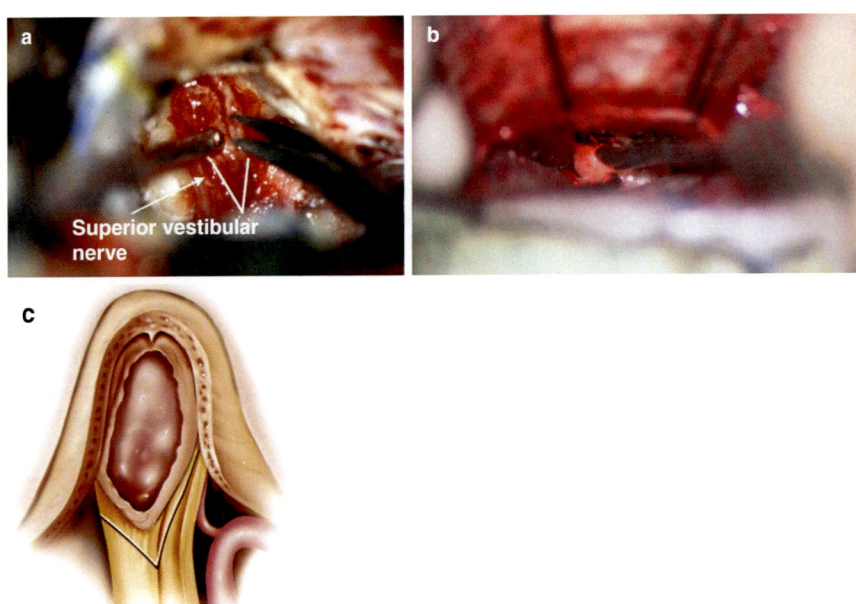

Fig. 5.7 V-cut technique. (**a**) The V-cut is shown here at the level of the medial end of IAC, resulting in a "V" shape between superior and inferior vestibular nerves. The lateral edge of the inferior vestibular nerve should be kept intact in order to protect the cochlear nerve for HP. As both the nerves lie in the same plane, the edge of the inferior vestibular nerve also helps to establish the dissection plane between the tumor and the cochlear nerve. In this figure, the V-cut is performed through microscissors. (**b**) In this figure, laser is used to perform the V-cut. (**c**) Diagram illustrating how the a surgical plane underneath the capsule and between the vestibular nerves is individuated by V-cut (*Fig. (c) reprinted from T. Fukushima, A. Friedman, L. Mastronardi, T. Sameshima, Fukushima's Microanatomy and Dissection of the Temporal Bone – Second Edition, 2007, with permission from AF-Neurovideo, Inc.*)

Debulking of the center of the tumor is important in VS surgery. The tumor should be removed piece by piece from the inside. The outer shell formed by the residual tumor should be gradually made as thin as possible (2–3-mm thick) in order to elevate the tumor capsule from the neurovascular structures without injuries. Soft and small tumors are enucleated with microdissectors and suction. However, sharp scissors are the tool of choice for gutting firmer tumors. The ultrasonic aspirator can be used for large, hard, or fibrous tumors [2, 11] (ideal setting: power 50, suction 5, irrigation 5).

5.7   Removal of the Tumor in the IAC

The portion of the tumor within the IAC must be removed piecemeal. The McElveen-Hitzelberger neural dissector (Fig. 5.8c) is the most appropriate instrument to use to dissect the tumor away from the fundus of the IAC [11]. This instrument can be used to separate the end of the tumor from the adjacent nerves, to cut the tumor from its

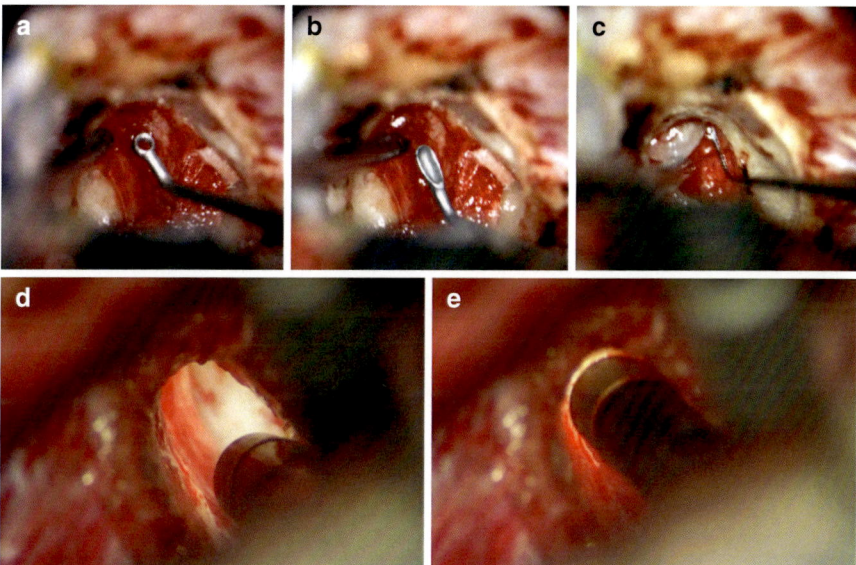

Fig. 5.8 Dissection and removal of VS away from the fundus of IAC. (**a**) 1-mm ring curette. (**b**) 1-mm cup curette. (**c**) Hitzelberger-McElveen knife. Tumor removal from the fundus of IAC must be performed piecemeal because this is the so-called blind region of the surgical field, and care should be taken not to injure labyrinthine structure, which would eventually preclude HP, and to avoid cranial nerves avulsion. Endoscopic-assisted surgery has been advocated as a means to implement microsurgical technique in the completion of tumor removal. (**d, e**) Insertion of flexible endoscope into the IAC (*Figs. (**d, e**) reprinted from World Neurosurgery, 115, Francesco Corrivetti, Guglielmo Cacciotti, Carlo Giacobbo Scavo, Raffaelino Roperto, Luciano Mastronardi, Flexible Endoscopic-Assisted Microsurgical Radical Resection of Intracanalicular Vestibular Schwannomas by a Retrosigmoid Approach: Operative Technique, Pages No. 229.233, 2018, with permission from Elsevier*)

nerve of origin, and to fragment the tumor. The use of bipolar coagulation must be minimized as the electrical current from the bipolar probes within the IAC may damage the N VII or cochlear nerve or the internal auditory artery (IAA), which is important for hearing preservation [1, 2, 8–11, 14].

Microsurgical resection allows a wide visual field over the surgical field; however, the complete resection of the portion of VS penetrating deeply into fundus of IAC is difficult under a straight microscopic view. This problem can be overcome with endoscopic devices, which allow total gross removal in most cases and provide satisfactory outcomes in terms of facial nerve and hearing preservation [18, 19]. Kumon et al. [20] did not find any significant differences in HP, N VII function, and tumor recurrence rates between purely microsurgical and endoscopic-assisted procedures. Instead, overall tumor resection was significantly higher when endoscope had been used, especially for VSs located beyond the mid-portion of the IAC. Insertion of 30-degree or 70-degree angled rigid endoscope enables to visualize the entire IAC, down to its fundus, and to identify residual tumor. As a result, the lesion can be removed completely under endoscopic guidance. Turek et al. [19] also used the endoscope to verify the integrity of the mastoid and to confirm that all opened air cells (if any) had been closed appropriately. Such approach allowed them to use only wax, rather than recently harvested muscle and glue, which is known to impair scar tissue formation in close vicinity of N VII and cochlear nerve and can imitate a residual tumor tissue on a follow-up scans.

Flexible endoscopes represent a step forward in endoscopic-assisted microsurgery. In a series of three cases of intracanalicular VSs (ICVSs), Corrivetti, Mastronardi et al. [21] used a 4-mm flexible video endoscope (4 mm × 65 cm, Karl Storz, Inc.) at the end of microsurgical resection. The endoscope was introduced under both microscopic and endoscopic visualization to prevent injury to CPA structures, and the endoscopic tip was oriented into the IAC in order to detect tumor residue hiding in the deeper portion of IAC. If residual tumor was identified, microsurgical resection was pursued, and further endoscopic controls were repeated until complete tumor removal. The authors' conclusion was that flexible endoscope appears to be particularly suitable for the surgical management of the ICVS due to the dimensions (4 mm × 65 cm, Karl Storz, Inc.) and the possibility to orient the endoscopic tip into the IAC, therefore obtaining an optimal visualization of the fundus (Fig. 5.8d, e).

5.8   Dissection of Capsule from the Brainstem and Cranial Nerves

Following tumor debulking, the remaining tumor capsule is removed with microsurgical tools. With standard microsurgical instruments (sharp dissectors, sickle knife, McElveen knife, straight and curved microscissors, ring and cup curettes), the tumor is separated from the brainstem and cranial nerves (Fig. 5.9) during continuous N VII and—when HP is attempted—cochlear nerve monitoring [2, 11].

Dissection of the functioning cochlear nerve requires extremely gentle microneurosurgical technique. The surgeon is able to identify the thin cochlear nerve as the yellowish-whitish band at the caudal aspect of the tumor. The proximal facial nerve

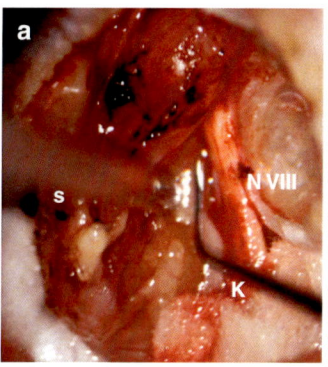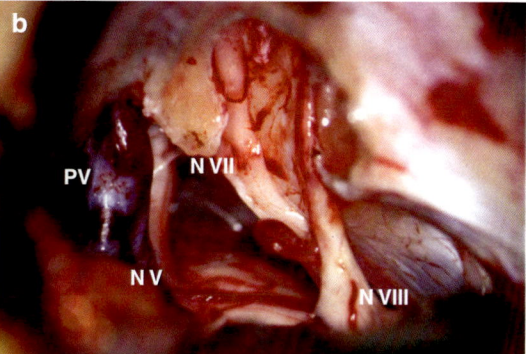

Fig. 5.9 Dissection of the capsule and final result. (**a**) Dissection of the capsule at the level of the internal acoustic meatus. *S* tapered tear drop suction, *K* Hitzelberger-McElveen knife, *N VIII* cochlear nerve. (**b**) Final view after hemostasis. *PV* petrosal vein, *N V* trigeminal nerve, *N VII* facial nerve, *N VIII* cochlear nerve

is always very white; however, the cochlear nerve is slightly yellowish because of the neural myelin. If HP is attempted, the surgeon cannot touch, push, or manipulate any of the cochlear nerve. It is very important to identify the true tumor capsule of the VS and elevate it gently and sharply while protecting the flattened cochlear nerve with Surgicel powder, Gelfoam, and supermicro cottonoids of 1–2 mm of size. If any decrease of the wave V is detected on the BAER, the surgeon must stop dissection, irrigate, and wait for recovery. Under the flocculus, the proximal remnant of the cochlear nerve is identified as a band running distally toward the IAC fundus.

The compressed nerves must be sharply dissected away from the tumor in a medial-to-lateral direction using thin blade microscissors or a sickle knife. This "sharp dissection" procedure, which avoids putting traction on the nerves, is very important for preserving nerve function. All strands of the arachnoid tethering the tumor should be cut sharply with microscissors rather than pulled. Firmly adherent tumor capsule should be thinned until it is transparent and supple rather than peeled away from the compressed nerves. In such cases, the surgeon should consider leaving 1 mm or 2 mm thickness of the thinned tumor capsule on the nerve [2].

Brief spurts of current rather than prolonged coagulation should be used for hemostasis because coagulation risks current spread and damage to the surrounding nerves. It is a good practice to use a cotton patty to isolate the nerves from the tips of the bipolar cautery forceps [2, 11]. Bone wall of IAC is covered with wax to occlude air cells and avoid CSF leaks [22], and canal is plugged with small pieces of muscle.

5.9   Closure

The autologous pericranium graft is inserted into the defect as an underlay hourglass-shaped plug [13]. For a successful result, the graft harvested has to be slightly larger than the dural defect, in order to have its edges under the dural plane. It is then fixed under operative microscope magnification with separated stitches (with an "inside-to-outside" direction) to the dura mater, using a 3-0 running silk [13]. After that, the inserted patch is augmented with one layer of absorbable hemostats (Fibrillar Surgicel, Ethicon, J and J, Somerville, New Jersey, USA), with small pieces of surgical patch (TachoSil®, Takeda, Japan), and with a dural sealant (DuraSeal, Covidien LLC, Mansfield, Massachusetts, or Tisseel, Baxter, Deerfield, Illinois, USA). The so-called surgical patch (TachoSil®, Takeda, Japan) combines the bioactive mechanism of action of fibrinogen and thrombin, with the mechanical support of a collagen patch. On contact with blood or other fluids, the coagulation factors react to form a fibrin clot that sticks the surgical patch to the tissue surface, producing an air- and liquid-tight seal in few minutes, providing protection against postoperative rebleedings and leaks [13].

In all cases, the previously removed autologous bone or fitted titanium net is placed on the bony defect with dedicated miniscrews, and HydroSet™ bone cement (Stryker Inc., Kalamazoo, MI) is injected to fill the remaining space [13]. The wound is finally closed in layers (Fig. 5.10).

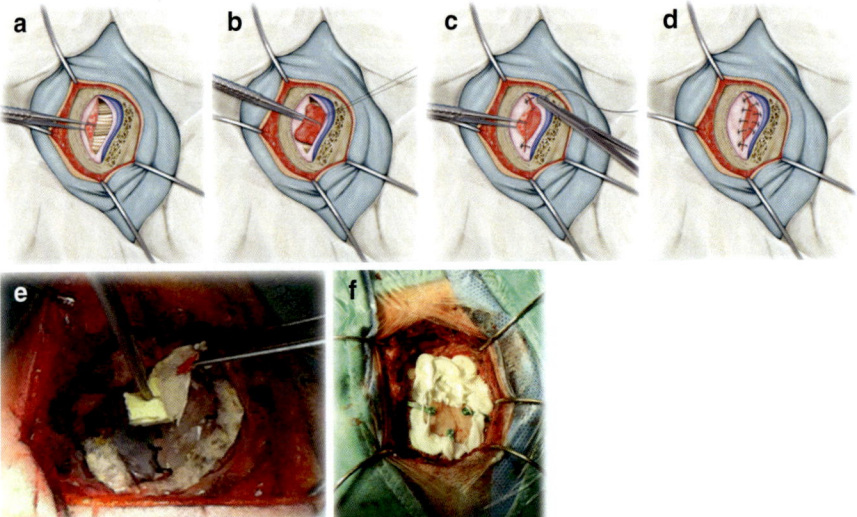

Fig. 5.10 Intraoperative picture of a typical "keyhole" retrosigmoid closure. (**a**) Step-by-step procedure. The local harvested pericranium is inserted under the dural plane (**b**) with "inside-to-outside" direction (**c**) and is stitched (**d**). TachoSil® is then applied to the bone defect in order to obtain tight sealing (**e**), and the bone is finally positioned back, fixed with miniscrews (Lorenz, Biomet Microfixation, Jacksonville, Florida, USA) and surrounded by filling injection of HydroSet™ bone cement (Stryker Inc., Kalamazoo, MI) (**f**). (*Fig. (a) reprinted from Surgical Neurology International, 7:25, Luciano Mastronardi, Guglielmo Cacciotti, Franco Caputi, Raffaelino Roperto, Maria Pia Tonelli, Ettore Carpineta, Takanori Fukushima, Underlay hourglass-shaped autologous pericranium duraplasty in "key-hole" retrosigmoid approach surgery: Technical report, 2016, from Medknow under Creative Commons BY copyright license*)

References

1. Samii M, Gerganov V, Samii A. Improved preservation of hearing and facial nerve function in vestibular schwannoma surgery via the retrosigmoid approach in a series of 200 patients. J Neurosurg. 2006;105(4):527–35.
2. Wanibuchi M, Fukushima T, Friedman AH, Watanabe K, Akiyama Y, Mikami T, et al. Hearing preservation surgery for vestibular schwannomas via the retrosigmoid transmeatal approach: surgical tips. Neurosurg Rev. 2014;37(3):431–44; discussion 44.
3. Scheller C, Wienke A, Tatagiba M, Gharabaghi A, Ramina KF, Ganslandt O, et al. Stability of hearing preservation and regeneration capacity of the cochlear nerve following vestibular schwannoma surgery via a retrosigmoid approach. J Neurosurg. 2016;125(5):1277–82.
4. Peng KA, Wilkinson EP. Optimal outcomes for hearing preservation in the management of small vestibular schwannomas. J Laryngol Otol. 2016;130(7):606–10.
5. Satar B, Yetiser S, Ozkaptan Y. Impact of tumor size on hearing outcome and facial function with the middle fossa approach for acoustic neuroma: a meta-analytic study. Acta Otolaryngol. 2003;123(4):499–505.
6. Anaizi AN, DiNapoli VV, Pensak M, Theodosopoulos PV. Small vestibular schwannomas: does surgery remain a viable treatment option? J Neurol Surg B Skull Base. 2016;77(3):212–8.
7. Sameshima T, Fukushima T, McElveen JT, Friedman AH. Critical assessment of operative approaches for hearing preservation in small acoustic neuroma surgery: retrosigmoid vs middle fossa approach. Neurosurgery. 2010;67(3):640–4; discussion 4–5.
8. Mastronardi L, Cacciotti G, Roperto R, Tonelli MP, Carpineta E. How I do it: the role of flexible hand-held 2μ-thulium laser fiber in microsurgical removal of acoustic neuromas. J Neurol Surg B Skull Base. 2017;78(4):301–7.
9. Mastronardi L, Cacciotti G, Scipio ED, Parziale G, Roperto R, Tonelli MP, et al. Safety and usefulness of flexible hand-held laser fibers in microsurgical removal of acoustic neuromas (vestibular schwannomas). Clin Neurol Neurosurg. 2016;145:35–40.
10. Mastronardi L, Di Scipio E, Cacciotti G, Roperto R. Vestibular schwannoma and hearing preservation: usefulness of level specific CE-Chirp ABR monitoring. A retrospective study on 25 cases with preoperative socially useful hearing. Clin Neurol Neurosurg. 2018;165:108–15.
11. Sameshima T. Fukushima's microanatomy and dissection of the temporal bone. 2nd ed. In: Sameshima T, editor. Raleigh: AF-Neurovideo, Inc.; 2007. 115 p.
12. Sameshima T, Mastronardi L, Friedman AH, Fukushima T. Microanatomy and dissection of temporal bone for surgery of acoustic neuroma and petroclival meningioma. 2nd ed. Raleigh: AF Neurovideo, Inc.; 2007.
13. Mastronardi L, Cacciotti G, Caputi F, Roperto R, Tonelli MP, Carpineta E, et al. Underlay hourglass-shaped autologous pericranium duraplasty in "key-hole" retrosigmoid approach surgery: technical report. Surg Neurol Int. 2016;7:25.
14. Tatagiba M, Roser F, Schuhmann MU, Ebner FH. Vestibular schwannoma surgery via the retrosigmoid transmeatal approach. Acta Neurochir. 2014;156(2):421–5; discussion 5.
15. Campero A, Martins C, Rhoton A, Tatagiba M. Dural landmark to locate the internal auditory canal in large and giant vestibular schwannomas: the Tübingen line. Neurosurgery. 2011;69(1 Suppl Operative):ons99–102; discussion ons102.
16. Tatagiba MS, Evangelista-Zamora R, Lieber S. Mobilization of the anterior inferior cerebellar artery when firmly adherent to the petrous dura mater-A technical nuance in retromastoid transmeatal vestibular schwannoma surgery: 3-dimensional operative video. Oper Neurosurg (Hagerstown). 2018;15(5):E58–9.
17. Ebner FH, Kleiter M, Danz S, Ernemann U, Hirt B, Löwenheim H, et al. Topographic changes in petrous bone anatomy in the presence of a vestibular schwannoma and implications for the retrosigmoid transmeatal approach. Neurosurgery. 2014;10(Suppl 3):481–6.
18. Tatagiba MS, Roser F, Hirt B, Ebner FH. The retrosigmoid endoscopic approach for cerebellopontine-angle tumors and microvascular decompression. World Neurosurg. 2014;82(6 Suppl):S171–6.

19. Turek G, Cotúa C, Zamora RE, Tatagiba M. Endoscopic assistance in retrosigmoid transmeatal approach to intracanalicular vestibular schwannomas—an alternative for middle fossa approach. Technical note. Neurol Neurochir Pol. 2017;51(2):111–5.
20. Kumon Y, Kohno S, Ohue S, Watanabe H, Inoue A, Iwata S, et al. Usefulness of endoscope-assisted microsurgery for removal of vestibular schwannomas. J Neurol Surg B Skull Base. 2012;73(1):42–7.
21. Corrivetti F, Cacciotti G, Giacobbo Scavo C, Roperto R, Mastronardi L. Flexible endoscopic-assisted microsurgical radical resection of intracanalicular vestibular schwannomas by a retrosigmoid approach: operative technique. World Neurosurg. 2018;115:229–33.
22. Nonaka Y, Fukushima T, Watanabe K, Friedman AH, Sampson JH, Mcelveen JT, et al. Contemporary surgical management of vestibular schwannomas: analysis of complications and lessons learned over the past decade. Neurosurgery. 2013;72(2 Suppl Operative):ons103–15; discussion ons15.

Translabyrinthine Approach

6

Luciano Mastronardi, Alberto Campione,
Guglielmo Cacciotti, Raffaelino Roperto,
Carlo Giacobbo Scavo, Ali Zomorodi,
and Takanori Fukushima

The translabyrinthine approach consists of an extradural procedure that gives access to the internal auditory canal (IAC) through mastoidectomy and drilling of semicircular canals and vestibule (Fig. 6.1). Due to iatrogenic damage to the membranous labyrinth, postoperative hearing loss is expected so that this approach is best indicated for preoperatively already deaf patients. Episodic cases of modest long-term hearing preservation have been reported after translabyrinthine approach [1, 2]. Diverse hypotheses have been proposed to explain such unusual functional outcome, the anatomical preservation of the vestibule being one of the most credited [1], but further studies are needed to determine the exact mechanisms. A rate as high as 27% has been reported as regards the functional preservation of the cochlear nerve. Kiyomizu et al. [3] found that smaller tumors not extending into the fundus of the IAC had a significantly higher correlation with postoperative positive response of the cochlear nerve to electrical stimulation test.

Advocates of this approach cite early identification of the facial nerve in the IAC and minimal brain or cerebellar retraction, which allows for removal of vestibular schwannomas (VSs) of any size. Springborg et al. [4] reported the results of outcome measures evaluation in 1244 patients operated on by translabyrinthine approach over 33 years. The rate of total tumor resection was 84% with a good facial outcome (House-Brackmann grade 1 or 2) in 70% of patients; a rate as high as 14% of CSF leakage was reported. Lanman et al. [5] and Zhang et al. [6] concentrated on large VSs (≥3 cm in largest diameter) and reported a total tumor removal

L. Mastronardi (✉) · A. Campione · G. Cacciotti · R. Roperto · C. Giacobbo Scavo
Department of Neurosurgery, San Filippo Neri Hospital—ASLRoma1, Rome, Italy
e-mail: mastro@tin.it

A. Zomorodi · T. Fukushima
Division of Neurosurgery, Duke University Medical Center, Carolina Neuroscience Institute,
Raleigh, NC, USA
e-mail: ali.zomorodi@duke.edu; Fukushima@carolinaneuroscience.com

© Springer Nature Switzerland AG 2019
L. Mastronardi et al. (eds.), *Advances in Vestibular Schwannoma Microneurosurgery*, https://doi.org/10.1007/978-3-030-03167-1_6

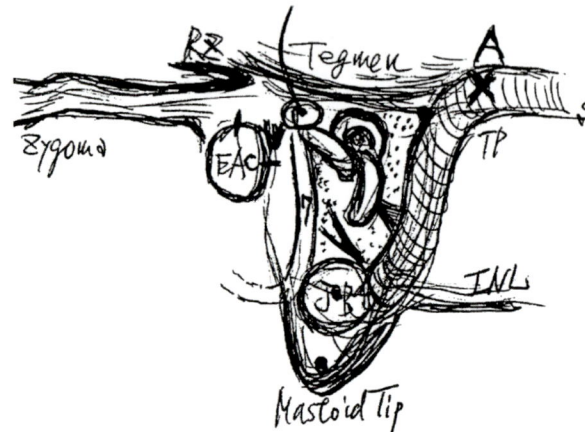

rate of 96.3% and 89.6%, respectively, despite the negative influence of greater tumor dimensions. The facial nerve was preserved anatomically in 93.7% and in 87.8% of the patients, respectively. CSF leakage that required surgical repair occurred in only 1.1% of the patients in the study by Lanman et al. [5]; Zhang et al. [6] reported a higher rate, 7%.

6.1 Positioning, Incision, and Bony Landmarks

The patient is placed in lateral Fukushima position or in supine position with the head tilted contralaterally to the tumor. A postauricular C-shaped incision through the galea is made, 2 cm behind the postauricular crease. However, in case of large tumors, a wider exposure of retrosigmoid dura is needed to allow access, and therefore the incision should be performed more posteriorly [7]. The incision extends from the mastoid tip and curves forward to end just above the pinna (or midpoint of supramastoid crest) [8] (Fig. 6.2).

The scalp is elevated by sharply dissecting the subgaleal connective tissue which spans the galea and the underlying pericranium. The pericranium is contiguous with the temporalis fascia above and the fascia overlying the sternocleidomastoid muscle below. A second incision is made in this deep layer composed of temporalis fascia and muscle, periosteum, and sternocleidomastoid fascia to fashion a musculofascial flap that is important in obtaining a watertight, cosmetic closure. Indeed, temporalis muscle is harvested from underneath the temporalis fascia and saved for packing of the eustachian tube and middle ear in later stages of the procedure [7]. The two flaps are elevated anteriorly to reveal the posterior lip of the external auditory canal, the spine of Henle, and the root of the zygoma (posterior point) [8] (Fig. 6.3). At this point, Fukushima's outer mastoid triangle is exposed; its landmarks are the asterion posteriorly, the root of the zygoma anteriorly, and the mastoid tip inferiorly (Fig. 6.3).

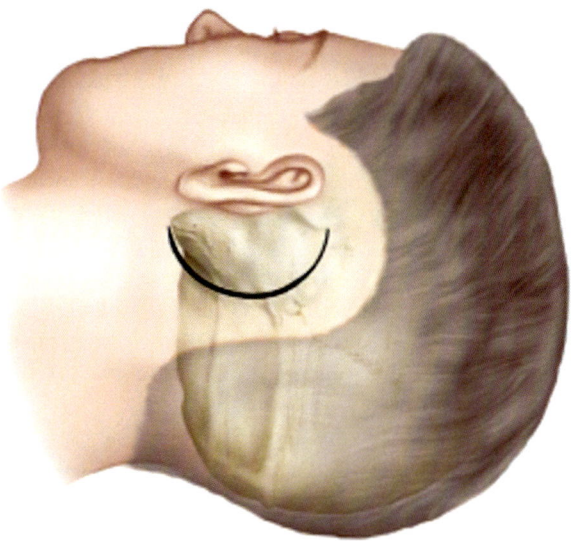

Fig. 6.2 C-shaped incision for retrolabyrinthine approach. (*Reprinted and modified from T. Fukushima, A. Friedman, L. Mastronardi, T. Sameshima, Fukushima's Microanatomy and Dissection of the Temporal Bone – Second Edition, 2007, with permission from AF-Neurovideo, Inc.*)

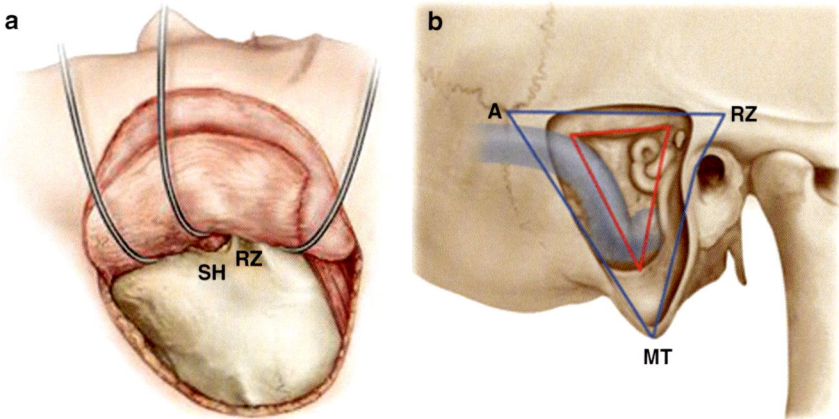

Fig. 6.3 (**a**) Anterior elevation of superficial and deep musculofascial flaps. *SH* spine of Henle, *RZ* root of zygoma. (**b**) Exposure of Fukushima's outer mastoid triangle, outlined in blue. *A* asterion, *RZ* root of the zygoma, *MT* mastoid tip. (*Reprinted and modified from T. Fukushima, A. Friedman, L. Mastronardi, T. Sameshima, Fukushima's Microanatomy and Dissection of the Temporal Bone – Second Edition, 2007, with permission from AF-Neurovideo, Inc.*)

6.2 Mastoidectomy

Using the high-speed drill with a large cutting burr (5–6 mm) and continuous irrigation, the cortex over the mastoid bone is removed. It is helpful to first outline the boundaries of bone to be removed using the drill. The anterior border is a slightly curved line, extending from the top of the external auditory meatus to the mastoid tip. The superior margin is along a line roughly perpendicular to the first, extending from the root of zygoma (posterior point) to the region of the asterion [9]. These two lines form a skewed "T" that defines the anterior and superior margins of the mastoidectomy—as outlined by the superior and anterior sides of Fukushima's outer mastoid triangle (Fig. 6.4). The junction of these two lines generally marks the surface projection of the region of the mastoid antrum and the lateral semicircular canal [8].

The bone cortex is removed within the boundaries of these lines, working anterior to posterior and superior to inferior. As the cortical bone is removed, air cells will be encountered. Posteriorly, over the sigmoid sinus, the bone will remain compact. In order to provide maximum exposure, wide cortical removal with saucerization should be performed prior to deeper penetration. Gentle, brushlike strokes with the drill will reveal the compact bone of the sigmoid sinus. A dominant sigmoid sinus may be the source of dramatic bleeding during the early stages of drilling; a preoperative angiographic study is therefore advisable so as to prevent excessively aggressive skeletonization of this structure. In addition, if a posterior displacement of the sigmoid sinus is needed to enlarge the surgical corridor, a small sigmoid sinus lowers the risk of cranial venous hypertension [8].

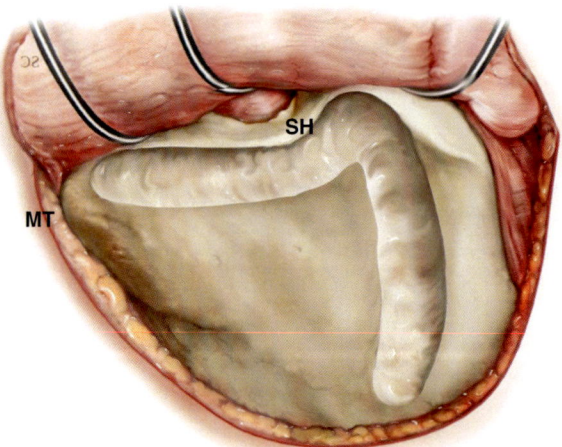

Fig. 6.4 First step of mastoidectomy. The boundaries of bone drilling are outlined along with the regional landmarks. *SH* spine of Henle, *MT* mastoid tip. (*Reprinted and modified from T. Fukushima, A. Friedman, L. Mastronardi, T. Sameshima, Fukushima's Microanatomy and Dissection of the Temporal Bone – Second Edition, 2007, with permission from AF-Neurovideo, Inc.*)

Bone removal proceeds 1 cm behind the sigmoid sinus, maintaining a uniform depth as it is exposed. Once it has been skeletonized, the mastoid air cells are removed anteriorly and superiorly to expose the middle fossa dura (temporal tegmen). Moving anteriorly, the air cells will be removed to expose the compact bone of the bony labyrinth. The key landmark in this area is the mastoid antrum, which defines the anterior limit of bony removal and locates the lateral semicircular canal (LSC) [9]. Maintaining the same relative depth, air cell removal proceeds inferiorly. As air cells are removed from the mastoid tip region, the digastric ridge will be encountered (Fig. 6.5). The digastric groove is an important landmark for defining the exit of the facial nerve from the fallopian canal through the stylomastoid foramen [9]. The stylomastoid foramen lies just medial to the anterior limit of the digastric ridge [8].

At this point, the middle fossa dura, presigmoid dura, and the sinodural angle have been skeletonized. The technique of "eggshelling"—i.e., removing bone until a thin shell is left, which may be removed with a dissector—is practiced to avoid damage to the dura and the venous structures. For maximal exposure in the retrolabyrinthine approach, the posterolateral portion of the bony labyrinth must be completely defined. With a medium diamond burr (2–3 mm), the small air cells surrounding the labyrinth are removed. The ridge covering the LSC is first identified as the antrum is opened. The tympanic segment of the facial nerve will be located parallel and 1–2 mm anterior to the lateral semicircular canal at this point (Fig. 6.6). Proceeding inferiorly, the descending segment of the facial nerve can be traced as a pink line running from the LSC to the digastric ridge (Fig. 6.6).

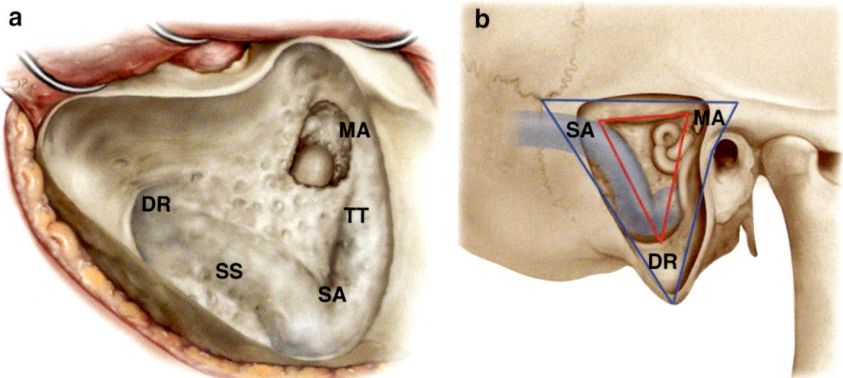

Fig. 6.5 Second step of the mastoidectomy. (**a**) The mastoid cells have been drilled and the sigmoid sinus has been eggshelled. The dura mater of posterior fossa (temporal tegmen or tegmen tympani) has been exposed along with the mastoid antrum. The inferior limit of such saucerization is the digastric ridge. *MA* mastoid antrum, *TT* temporal tegmen, *SA* sinodural angle (located between dura covering sigmoid sinus and temporal tegmen). *SS* sigmoid sinus, *DR* digastric ridge. (**b**) Fukushima's inner mastoid triangle is here outlined in red and corresponds to the surgical field of interest at this step of mastoidectomy. *SA* sinodural angle, *MA* mastoid antrum, *DR* digastric ridge. (*Reprinted and modified from T. Fukushima, A. Friedman, L. Mastronardi, T. Sameshima, Fukushima's Microanatomy and Dissection of the Temporal Bone – Second Edition, 2007, with permission from AF-Neurovideo, Inc.*)

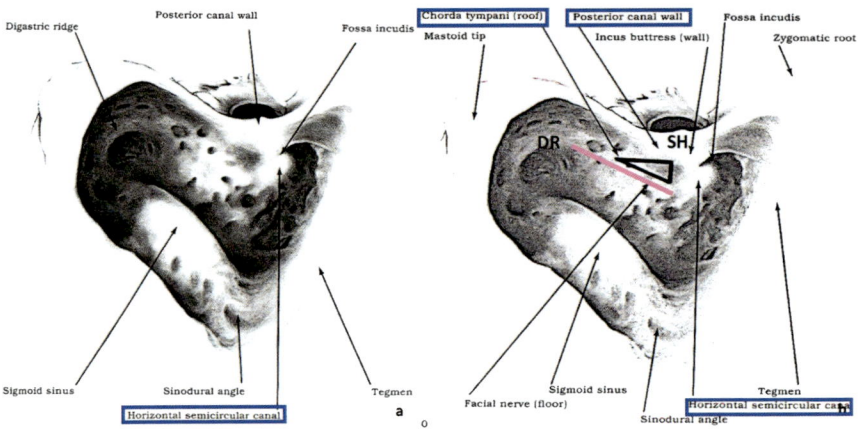

Fig. 6.6 Third (**a**) and fourth (**b**) steps of the mastoidectomy. (**a**) The horizontal (or lateral) semicircular canal is located deeply to the mastoid antrum and serves as a landmark to trace the course of the tympanic segment of the facial nerve. (**b**) The facial nerve (genu and descending portion) is identified as the pink line (see in the figure) running from the horizontal (or lateral) semicircular canal to the digastric ridge. It can also be traced as the third side of a triangle whose superior side is marked by the horizontal semicircular canal; the anterior side is limited by the posterior canal wall and corresponds to the roof of the chorda tympani. The triangle so outlined is the facial recess. *SH* spine of Henle, *DR* digastric ridge

Anteriorly, approximately 12–15 mm medial to the external auditory meatus, lies the fallopian canal. Therefore, bone removal in the anterior direction at this level must be done with extreme care to avoid violating the fallopian canal. The facial nerve, which lies anterior to the labyrinthine structures, is carefully approached, again using the lateral semicircular canal as a landmark. The facial nerve is skeletonized by using the diamond burr from the external genu inferiorly to the stylomastoid foramen. Care is taken to preserve a thin shell of bone around the facial nerve for protection (Fig. 6.7). This maneuver must be done under constant, copious irrigation to dissipate heat from drill [8].

6.3 Retrolabyrinthine Mastoidectomy: Final Exposure

Drilling of labyrinthine structures is performed within the limits of the third and deepest mastoid triangle, MacEwen's (or suprameatal) triangle. This region is located within Fukushima's inner mastoid triangle, and its main landmark remains the mastoid antrum. The superior side is half the length of the line running from the mastoid antrum to the sinodural angle. The anterior limit of MacEwen's triangle is half the length of the line running from the mastoid antrum to the digastric ridge. The resulting posteroinferior limit of MacEwen's triangle is an oblique line passing through the posterior semicircular canal (Fig. 6.8).

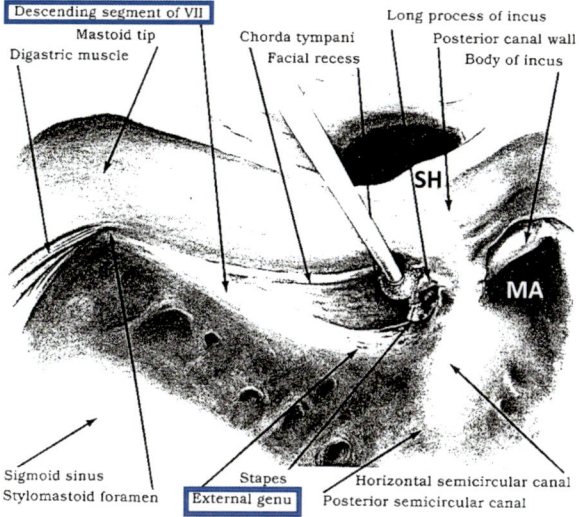

Descending segment of VII
Mastoid tip
Digastric muscle
Chorda tympani
Facial recess
Long process of incus
Posterior canal wall
Body of incus

SH

MA

Sigmoid sinus
Stylomastoid foramen
Stapes
External genu
Horizontal semicircular canal
Posterior semicircular canal

Fig. 6.7 Fifth step of the mastoidectomy. The facial nerve is eggshelled over its tympanic segment, genu, and descending segment; skeletonizing the anterior margin of the descending segments allows for the exposure of the facial recess (in the figure, it is being drilled). *SH* spine of Henle, *MA* mastoid antrum, *T* tympanic segment of the facial nerve, running parallel and 1–2 mm anterior to the horizontal (or lateral) semicircular canal

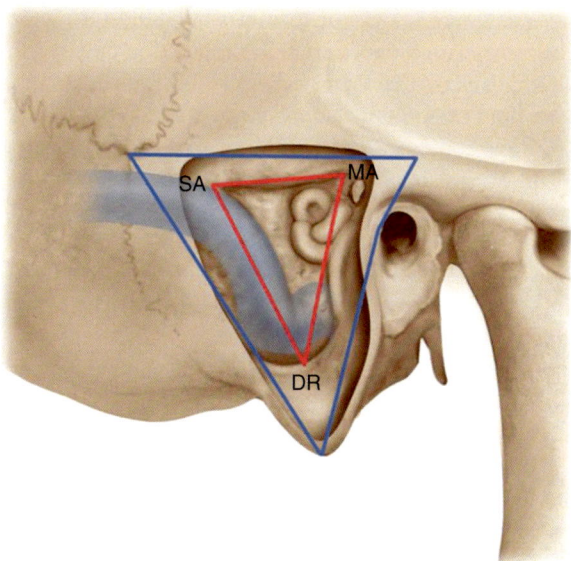

SA   MA

DR

Fig. 6.8 MacEwen's triangle. MacEwen's triangle marks the region to drill in order to expose the membranous labyrinth. *MA* mastoid antrum, *DR* digastric ridge, *SA* sinodural angle. (*Reprinted and modified from T. Fukushima, A. Friedman, L. Mastronardi, T. Sameshima, Fukushima's Microanatomy and Dissection of the Temporal Bone – Second Edition, 2007, with permission from AF-Neurovideo, Inc.*)

The exposure of the semicircular canals may be cumbersome in that the labyrinthine structures are fragile and may be damaged by excessive drilling; in order to avoid such inconvenience, gentle eggshelling should be performed until a transparent blue line can be seen in each of the canals. In addition, spatial disposition of such structures may pose problems of orientation within the surgical field. In this case, a schematic representation of the regional anatomy may be useful, as shown in Fig. 6.9.

Moving posteriorly, as the posterior semicircular canal (PSC) is defined, the retrofacial air cells are located just inferior to it and toward the jugular dome. These air cells are removed to skeletonize the jugular bulb [8] (Fig. 6.10).

6.4   Retrolabyrinthine Approach

After presigmoid dura exposure, the dura is incised parallel to the presigmoid region and the superior petrosal sinus. The dura is retracted anteriorly exposing VS and cranial nerve VII and VIII in the cerebellopontine angle (CPA). Frequently the lower cranial nerves can also be visualized [8] (Fig. 6.11). Once the dura has been opened, the cerebellum is protected with cottonoids. It is important to locate and preserve the arachnoid sheath that overlays the tumor capsule and separate it away from the parenchyma in order to stay in the appropriate plane. The posterior pole of the tumor is identified, and it is advised to put the monopolar probe over the tumor capsule in order to rule out a posterior course of the facial nerve. It is useful to go downward early in order to cut the arachnoid under the inferior pole of the tumor and then release the cerebrospinal fluid from the cerebellomedullary cistern. It gives significant cerebellar relaxation and avoids application of retractors. The tumor mass can then be debulked using ring curettes or ultrasonic aspiration, depending on the tumor consistency. Inside the CPA, the facial nerve is dissected from the tumor capsule in the medial to lateral direction. The continuous use of the stimulator helps to determine the trajectory of the nerve over the tumor periphery [9, 10].

6.5   Translabyrinthine Drilling and Opening of the IAC

The LSC and PSC are first opened with the drill. The amputated, or anterior, end of the LSC is carefully removed, bearing in mind the close relationship of the tympanic portion of the facial nerve. Preservation of the anterior wall of the lateral semicircular canal will protect the tympanic segment of the facial nerve. Removal of the superior segment of the posterior semicircular canal will expose the common crus which it shares with the superior semicircular canal (SSC). The SSC is then also opened by drilling superiorly and anteriorly [7, 9–11]. The amputated, or inferior, limb of the PSC is followed to the vestibule. Drilling in this area, lateral and inferior to the vestibule, will expose the vestibular aqueduct as it courses laterally toward the endolymphatic sac (Fig. 6.12). The vestibule is then opened by continuing to remove bone, following the common crus. The medial wall of the vestibule, which separates it from the IAC, is only a thin shell of bone and corresponds to the posterior border of the IAC [9] (Fig. 6.13).

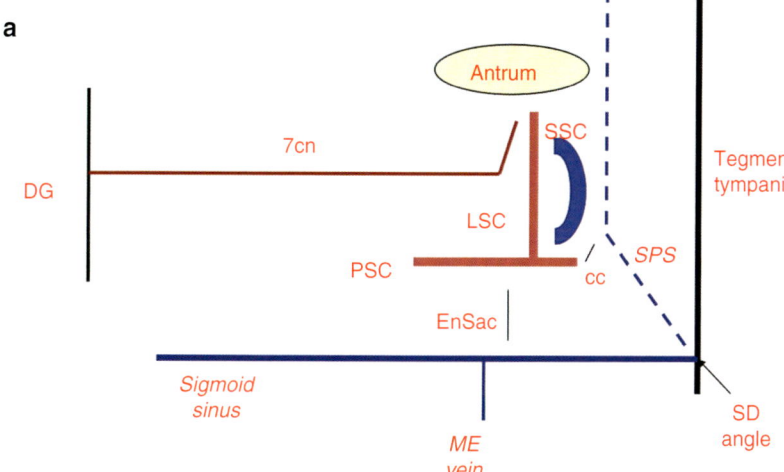

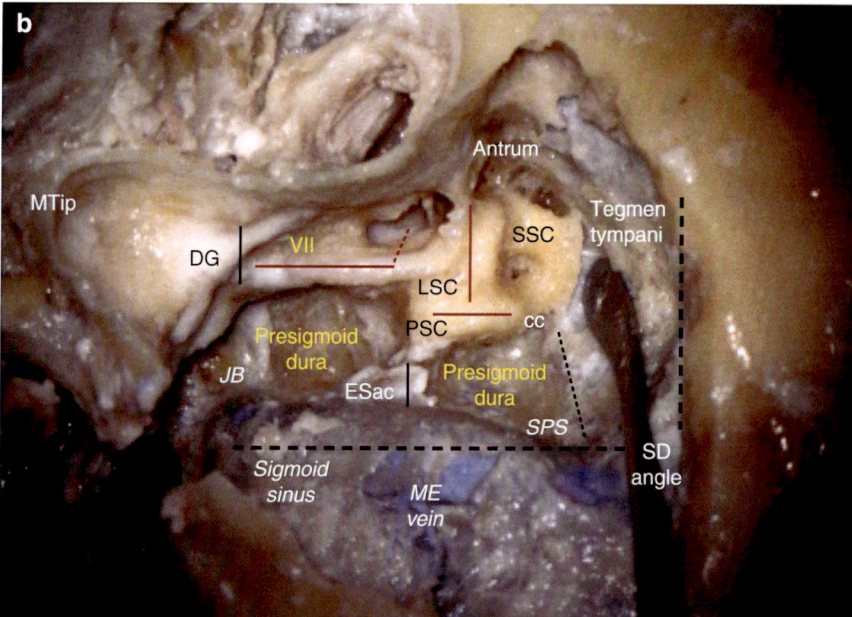

Fig. 6.9 Rule of perpendicular lines. (**a**) Schematic representation of regional anatomy arranged as multiple parallel and perpendicular lines. The central and main landmark is the antrum and the lateral semicircular canal (LSC) that becomes visible once the antrum is opened. Posterior and perpendicular to the LSC is the posterior semicircular canal (PSC). The superior semicircular canal (SSC) shares the common crus (cc) with the PSC. Further superiorly, the superior petrosal sinus (SPS) is encountered and then the middle fossa dura or tegmen tympani, which extends posteriorly until covering the sigmoid sinus and thus yielding the sinodural angle (SD angle). The sigmoid sinus is crossed perpendicularly both anteriorly and posteriorly by two structures, the endolymphatic sac (EnSac) and the mastoid emissary vein (ME vein), respectively. As regards the facial nerve (7cn), its tympanic segment is anteroinferior and approximately parallel to LSC. Running inferiorly, the 7cn is parallel to the sigmoid sinus and ends perpendicular at the level of the digastric ridge (DG). (**b**) The schematic representation is superimposed to a "surgical" photograph showing a left mastoidectomy. As a result of progressive eggshelling, the presigmoid dura has become apparent, and the sigmoid sinus has been skeletonized down to the jugular bulb (JB)

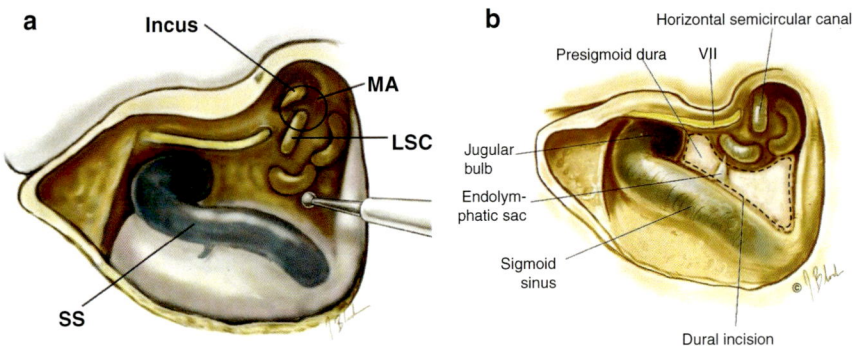

Fig. 6.10 (**a**) Jugular bulb exposure after drilling of the retrofacial air cells. (**b**) Presigmoid dura exposure and final result of the mastoidectomy. (*Reprinted and modified from T. Fukushima, A. Friedman, L. Mastronardi, T. Sameshima, Fukushima's Microanatomy and Dissection of the Temporal Bone – Second Edition, 2007, with permission from AF-Neurovideo, Inc.*)

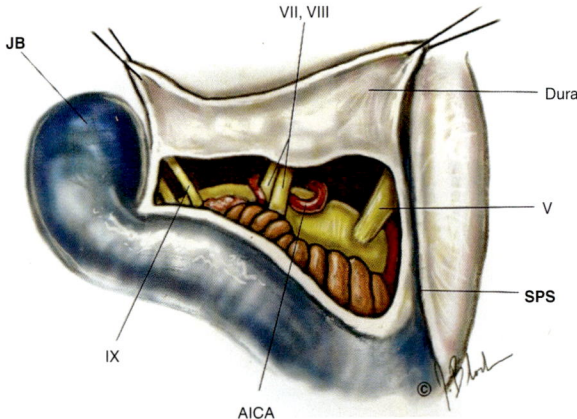

Fig. 6.11 Presigmoid dura opened, left side. It is worth noting how the brainstem and the cranial nerve are correctly and clearly visualized without the need for cerebellar retraction, which is one of the main advantages reported by the advocates of combined translabyrinthine and retrolabyrinthine approaches. *SPS* superior sagittal sinus, *AICA* anterior inferior cerebellar artery, *JB* jugular bulb. (*Reprinted and modified from T. Fukushima, A. Friedman, L. Mastronardi, T. Sameshima, Fukushima's Microanatomy and Dissection of the Temporal Bone – Second Edition, 2007, with permission from AF-Neurovideo, Inc.*)

The compact bone surrounding the IAC is defined by removing bone superior and inferior to the canal. It is important to remove the bone around the canal, superiorly and inferiorly, such that 180–270° [7, 10, 11] of the circumference of the canal is skeletonized. However, when large or giant VSs are encountered, neurovascular structures usually lie anteriorly to the tumor and are hidden from view. Blind separation of tumor from such structures should never be attempted. To overcome this problem, transapical extension (type I) of the translabyrinthine approach was designed. It consists of removing the bone around and anterior to the IAC between 300° and 320°, which provides direct visualization and allows surgical control of

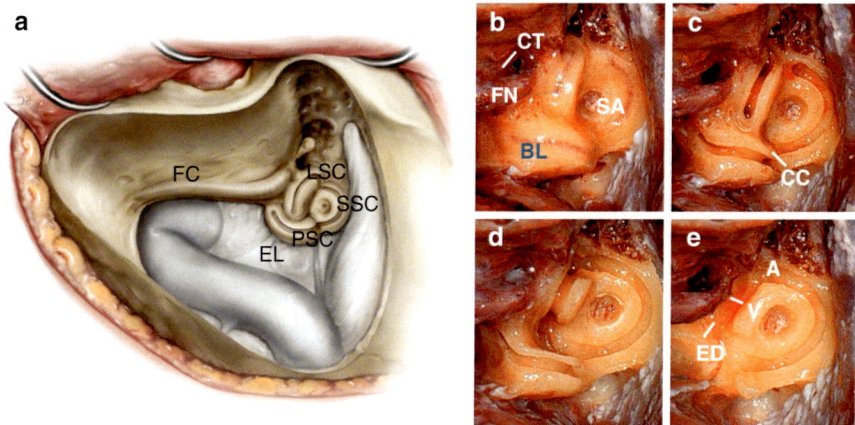

Fig. 6.12 Drilling of the semicircular canals. (**a**) Diagram of the surgical field after mastoidectomy and deep drilling. The semicircular canals are exposed. Note the close relation between lateral semicircular canal and fallopian canal, on one side, and between posterior semicircular canal and endolymphatic sac. *FC* fallopian canal, *LSC* lateral semicircular canal, *SSC* superior semicircular canal, *PSC* posterior semicircular canal, *EL* endolymphatic canal. (**b–d**) Photographs of progressive drilling of the semicircular canals. (**b**) *CT* chorda tympani, *FN* facial nerve running within the fallopian canal, *SA* subarcuate artery. (**c**) *CC* common crus. (**e**) *A* ampulla of the superior semicircular canal, *V* vestibule, *ED* endolymphatic duct. (*Figure (a) reprinted and modified from T. Fukushima, A. Friedman, L. Mastronardi, T. Sameshima, Fukushima's Microanatomy and Dissection of the Temporal Bone – Second Edition, 2007, with permission from AF-Neurovideo, Inc.*)

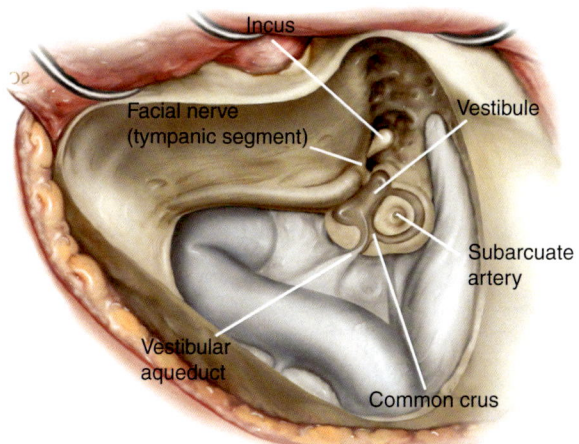

Fig. 6.13 Opening of the vestibule. (*Reprinted from T. Fukushima, A. Friedman, L. Mastronardi, T. Sameshima, Fukushima's Microanatomy and Dissection of the Temporal Bone – Second Edition, 2007, with permission from AF-Neurovideo, Inc.*)

areas that would otherwise be blind zones. The anterior boundaries of the tumor and neurovascular structures lying anteriorly become clearly visible, making dissection of the tumor capsule much safer [12–14].

The floor of the IAC is uncovered first by removing the bone that separates it from the jugular dome inferiorly [13]. If a high-riding jugular bulb is encountered, it is first skeletonized and then gently pushed inferiorly with the periosteum and held in place with bone wax [12–14]. Beginning in the region of the porus acusticus, the compact bone surrounding the canal is thinned with a small diamond burr until only a thin, repressible shell remains. As the drilling proceeds laterally, it should be remembered that the dura only covers the canal contents up to approximately two-thirds of the canal's length. As the bone is thinned at the lateral end of the canal, the transverse crest will be identified as a thin septum of bone separating the superior from inferior vestibular nerves [7, 9–11]. The paper-thin shell of bone in the region of the porus acusticus is removed first with a fine dissector, with the bone over the lateralmost end of the IAC saved for last (Fig. 6.14). The superior tip of the porus is generally the most difficult to manage because of the very close proximity of the facial nerve.

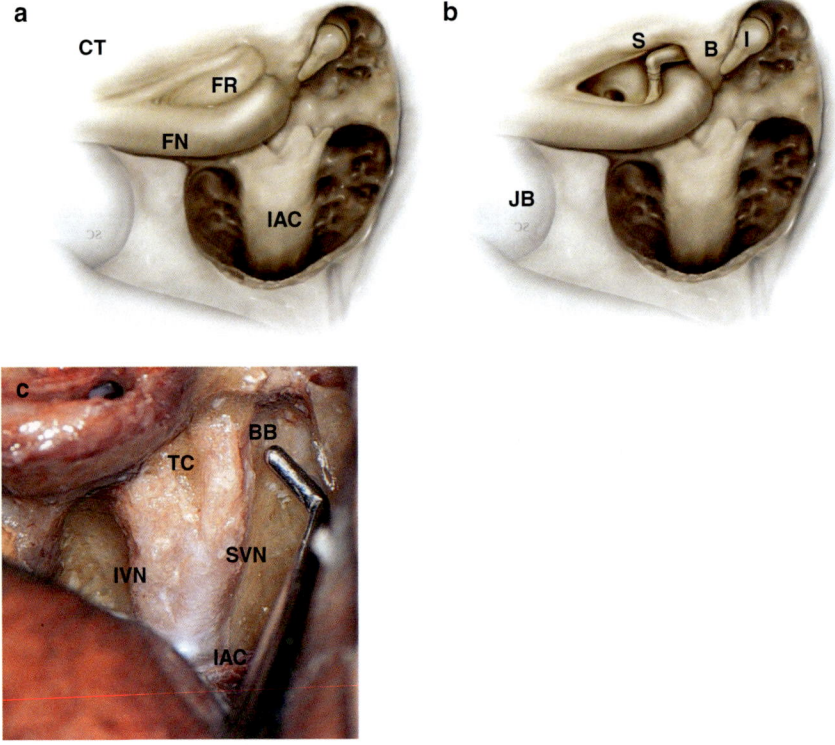

Fig. 6.14 Opening of the internal acoustic canal. (**a**) Diagram of the lateral end of the internal acoustic canal. *CT* chorda tympani, *FN* facial nerve, *FR* facial recess, *IAC* internal acoustic canal. (**b**) Diagram of the lateral end of the internal acoustic canal after drilling of the facial recess. Note the ossicular chain. *I* incus, *B* buttress, *S* stapes, *JB* jugular bulb. (**c**) The transverse crest and the Bill's bar have been recognized. Note that the transverse crest divides the superior and inferior vestibular nerves. Medially, the internal acoustic canal is covered by dura. *TC* transverse crest, *BB* Bill's bar, *IVN* inferior vestibular nerve, *SVN* superior vestibular nerve. (*Figures (**a**, **b**) reprinted and modified from T. Fukushima, A. Friedman, L. Mastronardi, T. Sameshima, Fukushima's Microanatomy and Dissection of the Temporal Bone – Second Edition, 2007, with permission from AF-Neurovideo, Inc.*)

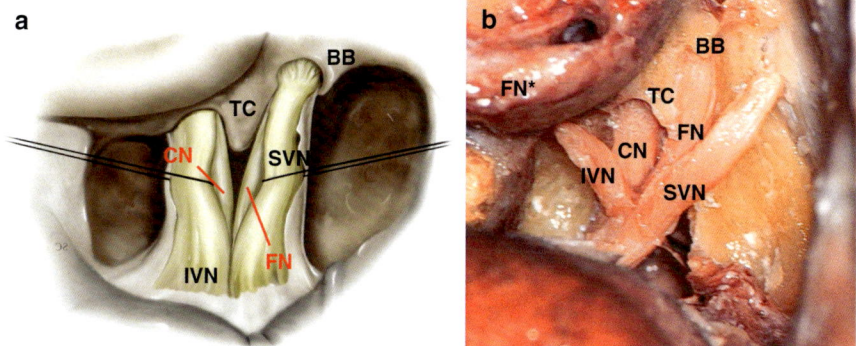

Fig. 6.15 Dissection of the contents within the internal auditory canal. (**a**) Diagram showing that the transverse crest interposes between the two vestibular nerves. The Bill's bar serves as a landmark to recognize the surgical plane of dissection between the superior vestibular nerve and the facial nerve, here emphasized in red as the cochlear nerve. *TC* transverse crest, *BB* Bill's bar, *CN* cochlear nerve, *SVN* superior vestibular nerve, *IVN* inferior vestibular nerve, *FN* facial nerve. (**b**) The contents of the internal auditory canal have been dissected from the facial nerve, which runs laterally and vertically in the fallopian canal (FN*). (*Figure (**a**) reprinted and modified from T. Fukushima, A. Friedman, L. Mastronardi, T. Sameshima, Fukushima's Microanatomy and Dissection of the Temporal Bone – Second Edition, 2007, with permission from AF-Neurovideo, Inc.*)

The fundus of the canal is exposed just medial to the vestibule, and the transverse crest that separates the vestibular nerves is identified. The dura of the IAC is then sharply opened using a #11 blade. The transverse crest—between superior and inferior vestibular nerves—and the Bill's bar (vertical crest), between superior vestibular nerve and facial nerve, are identified in the lateral aspects of the IAC with a right-angle pick. The superior vestibular nerve is gently displaced while palpating the Bill's bar to allow visualization of the facial nerve, whose certain identification is confirmed with bipolar stimulation at minimal settings (0.05 mA) [11]. A small right-angle hook is used to avulse the superior vestibular nerve laterally and expose the facial nerve. The inferior vestibular nerve and cochlear nerves are then separated laterally (Fig. 6.15). The contents of the IAC are dissected from the facial nerve in a lateral to medial fashion and then removed [7, 9–11]; the cochlear nerve may be preserved if hearing preservation is attempted.

6.6   Closure

The incus is removed, and the piece of temporalis muscle previously harvested is packed carefully through the epitympanum occluding the origin of the eustachian tube. The removal of the incus and the obliteration of the eustachian tube entrance by muscle reduce the possibility of CSF leakage. The dural incision is closed up to the internal auditory canal, and strips of autologous (abdominal) fat are placed in the gaps of the dura so as to seal the CSF space [8] and must be taken not to iatrogenically replace tumor mass with fat mass in the CPA [7]. In an eight-patient series, Liu et al. [15] performed a dural "sling" reconstruction technique using autologous fascia lata to repair presigmoid dural defects to avoid the risk of direct compression of the fat graft on the facial nerve and brainstem. In their study, the fascia lata was sewn to the edges of the presigmoid dural defect to create a sling to suspend the fat graft within the mastoidectomy hole.

The previously fashioned musculofascial flap is closed tightly over the adipose graft, and the postauricular incision is closed in two layers.

References

1. Tringali S, Bertholon P, Chelikh L, Jacquet C, Prades JM, Martin C. Hearing preservation after modified translabyrinthine approach performed to remove a vestibular schwannoma. Ann Otol Rhinol Laryngol. 2004;113(2):152–5.
2. Tringali S, Ferber-Viart C, Gallégo S, Dubreuil C. Hearing preservation after translabyrinthine approach performed to remove a large vestibular schwannoma. Eur Arch Otorhinolaryngol. 2009;266(1):147–50.
3. Kiyomizu K, Matsuda K, Nakayama M, Tono T, Matsuura K, Kawano H, et al. Preservation of the auditory nerve function after translabyrinthine removal of vestibular schwannoma. Auris Nasus Larynx. 2006;33(1):7–11.
4. Springborg JB, Fugleholm K, Poulsgaard L, Cayé-Thomasen P, Thomsen J, Stangerup SE. Outcome after translabyrinthine surgery for vestibular schwannomas: report on 1244 patients. J Neurol Surg B Skull Base. 2012;73(3):168–74.
5. Lanman TH, Brackmann DE, Hitselberger WE, Subin B. Report of 190 consecutive cases of large acoustic tumors (vestibular schwannoma) removed via the translabyrinthine approach. J Neurosurg. 1999;90(4):617–23.
6. Zhang Z, Wang Z, Huang Q, Yang J, Wu H. Removal of large or giant sporadic vestibular schwannomas via translabyrinthine approach: a report of 115 cases. ORL J Otorhinolaryngol Relat Spec. 2012;74(5):271–7.
7. Arriaga MA, Lin J. Translabyrinthine approach: indications, techniques, and results. Otolaryngol Clin North Am. 2012;45(2):399–415, ix.
8. Sameshima T, Mastronardi L, Friedman AH, Fukushima T. Microanatomy and dissection of temporal bone for surgery of acoustic neuroma and Petroclival meningioma. 2nd ed. Raleigh: AF Neurovideo, Inc.; 2007.
9. Roche PH, Pellet W, Moriyama T, Thomassin JM. Translabyrinthine approach for vestibular schwannomas: operative technique. Prog Neurol Surg. 2008;21:73–8.
10. Nickele CM, Akture E, Gubbels SP, Başkaya MK. A stepwise illustration of the translabyrinthine approach to a large cystic vestibular schwannoma. Neurosurg Focus. 2012;33(3):E11.
11. Bennett M, Haynes DS. Surgical approaches and complications in the removal of vestibular schwannomas. Otolaryngol Clin N Am. 2007;40(3):589–609, ix–x.
12. Angeli RD, Piccirillo E, Di Trapani G, Sequino G, Taibah A, Sanna M. Enlarged translabyrinthine approach with transapical extension in the management of giant vestibular schwannomas: personal experience and review of literature. Otol Neurotol. 2011;32(1):125–31.
13. Ben Ammar M, Piccirillo E, Topsakal V, Taibah A, Sanna M. Surgical results and technical refinements in translabyrinthine excision of vestibular schwannomas: the Gruppo Otologico experience. Neurosurgery. 2012;70(6):1481–91; discussion 91.

14. Jayashankar N, Morwani KP, Sankhla SK, Agrawal R. The enlarged translabyrinthine and transapical extension type I approach for large vestibular schwannomas. Indian J Otolaryngol Head Neck Surg. 2010;62(4):360–4.
15. Liu JK, Patel SK, Podolski AJ, Jyung RW. Fascial sling technique for dural reconstruction after translabyrinthine resection of acoustic neuroma: technical note. Neurosurg Focus. 2012;33(3):E17.

Video Clips of the Surgical Steps

7

Luciano Mastronardi, Alberto Campione,
Guglielmo Cacciotti, Raffaelino Roperto,
Carlo Giacobbo Scavo, Ali Zomorodi,
and Takanori Fukushima

7.1   Video 7.1: Keyhole Craniotomy

This is a case of right-sided vestibular schwannoma, grade T4a according to Samii's classification (MRI is shown). A 5-mm longitudinal groove runs at the posterior border of the mastoid body, exposing safely the sigmoid sinus (on the top side of the screen). Then, the groove continues downward along the inferior edge of the proposed bone flap (on the right side of the screen), as well as along the superior margin to expose the junction of the sigmoid sinus and the transverse sinus (at the top left corner of the screen; the drill is being used under constant irrigation and aspiration). A 4-mm extra-coarse diamond drill is used.

7.2   Video 7.2: Opening of the Dura

The dura is initially pierced inferiorly, and the arachnoid membrane of the lateral medullary cistern is opened for CSF aspiration (00:10). The opening of dura proceeds superiorly in a semicircular fashion so as to cover part of the cerebellar hemisphere in order to serve as a protective sheath during retraction (00:22). A rectangular nitrile sheet, previously cut out of a surgical glove, is placed on the cerebellar cortex so as to avoid direct contact with the pledget (00:36). The pledget absorbs CSF, thus allowing for further cerebellar relaxation.

Electronic Supplementary Material The online version of this chapter (https://doi.org/10.1007/978-3-030-03167-1_7) contains supplementary material, which is available to authorized users.

L. Mastronardi (✉) · A. Campione · G. Cacciotti · R. Roperto · C. Giacobbo Scavo
Department of Neurosurgery, San Filippo Neri Hospital—ASLRoma1, Rome, Italy
e-mail: mastro@tin.it

A. Zomorodi · T. Fukushima
Division of Neurosurgery, Duke University Medical Center, Carolina Neuroscience Institute, Raleigh, NC, USA
e-mail: ali.zomorodi@duke.edu; Fukushima@carolinaneuroscience.com

© Springer Nature Switzerland AG 2019   69
L. Mastronardi et al. (eds.), *Advances in Vestibular Schwannoma Microneurosurgery*, https://doi.org/10.1007/978-3-030-03167-1_7

7.3 Video 7.3: Identification of Facial Nerve

The cerebellum is being held posteriorly by a spatula so that insight is warranted into the cerebellopontine angle. The arachnoid adhesions between the tumor capsule and the brainstem are lysed with the aid of a microsurgical dissector (00:05). The proximal segment of the facial nerve is identified and stimulated with a monopolar probe (00:10); based on its proximal displacement pattern, the nerve is supposed to run anteriorly/anteroinferiorly to the tumor. Finally, cottonoids are positioned between the tumor capsule and the brainstem (00:25).

7.4 Video 7.4: V-cut with Thulium Laser

According to Fukushima's technique, a V-cut is performed on the dorsal surface of the tumor with thulium laser fiber to initiate the debulking phase of the procedure. The edges of the V-cut are traced juxtaposing the tip of the laser fiber to the tissue; no deep penetration into the tumor bulk is required at the moment.

7.5 Video 7.5: Opening of the Internal Auditory Canal

The dura posterior to the porus acusticus is incised by laser (handheld 2μ-thulium flexible laser fiber, RevoLix jr®, Lisa laser USA, Pleasanton, CA, USA) in an inverted "U" shape, with its base approximately at the fovea. The inverted U-shaped dural incision extends about 6–8 mm toward the fovea, and its base extends 2 mm above and 2 mm below the internal auditory canal (00:02). Dural dissection from the pyramid is performed with the aid of a sharp dog dissector (00:17). The canal is exposed with Sonopet Ultrasonic Aspirator (Stryker, Kalamazoo, MI); a bone tip is used under constant irrigation so as avoid nerve damage due to bone overheating (00:26). As long as internal auditory canal exposure proceeds, further dural detachment is performed with laser fiber, which is also used for canalicular dural incision (01:04).

7.6 Video 7.6: Microdissection in the Internal Auditory Canal

The microdissection in the internal auditory canal consists of a stepwise procedure of exposure, dissection, and piecemeal removal. Microscissors (in this case, Kamiyama type) are used to cut any adhesions or tumor tissue preventing the surgical exposure (00:08). The sharp dog dissector is used to isolate the tumor inside the internal auditory canal (00:14). Microscissors (00:18), Hitzelberger-McElveen knife (bullet tip) (00:22), and 1-mm ring curette (00.26) are used for tumor fragmentation and removal. Any small bleeding or floating tumor residue is aspirated (00:29). Further exposure of the canal is performed with Sonopet Ultrasonic Aspirator (Stryker, Kalamazoo, MI) (00.38), and microdissection proceeds again as previously described.

7.7   Video 7.7: Flexible Endoscopy Final Check in the Internal Auditory Canal

This is a case of left-sided intracanalicular vestibular schwannoma, grade T1 according to Samii's classification (MRI is shown). A retrosigmoid craniotomy is performed, and the dura mater is performed in a semicircular fashion (00:09). CSF is aspirated to allow for cerebellar relaxation, and the arachnoid adhesions between the tumor capsule and the brainstem are lysed with the aid of a microsurgical dissector (00:15). The dura posterior to the porus acusticus is incised by laser (hand-held 2μ-thulium flexible laser fiber, RevoLix jr®, Lisa laser USA, Pleasanton, CA, USA) and dissected from the pyramid with the aid of a sharp dog dissector (00:20). The internal auditory canal is exposed with 4-mm coarse diamond burr and Sonopet Ultrasonic Aspirator (Stryker, Kalamazoo, MI) (00:29). The canalicular dura mater is incised by laser fiber (00:40), and microdissection is performed by means of microscissors, dissectors, curettes, and micro alligator tumor forceps (00:52). Flexible endoscopic assistance is employed to visualize any tumor residue at the lateralmost end of the internal auditory canal (02:14). A 1-mm ring curette is used to remove the tumor remnant (02:30). Second endoscopic control (02:40) is performed to confirm total tumor resection (02:40).

7.8   Video 7.8: Final View

The vestibular schwannoma has been totally resected, and both facial nerve and brainstem decompressions have been obtained.

Results in a Personal Series of 160 Cases

8

Luciano Mastronardi, Alberto Campione,
Guglielmo Cacciotti, Raffaelino Roperto,
and Carlo Giacobbo Scavo

The aim of this chapter is to illustrate the results in a personal series of 160 cases of vestibular schwannoma (VS) treated by Prof. Luciano Mastronardi and the team of the Division of Neurosurgery, Department of Neurological Sciences, San Filippo Neri Hospital, Rome, Italy. The report has been divided into three main parts regarding the patients' preoperative status, the results of surgery, and the postoperative functional outcomes, respectively. In addition, surgical results and functional outcomes have been analyzed as stratified according to the intraoperative techniques/devices employed—i.e., flexible endoscope, 0.3% diluted papaverine, 2μ-thulium laser fiber, and hydroxyapatite (HAC) fluid cement.

Prof. Takanori Fukushima has a personal experience of more than 2200 cases treated over 40 years; however, his personal results will not be discussed hereafter.

8.1 Preoperative Data

One hundred sixty consecutive patients (74 women and 86 men) were operated at our institution between September 2010 and April 2018. Mean age was 49.9 years, and no statistically significant difference was reported between women's and men's mean ages (women's mean age, 51.5 years; men's mean age, 48.5 years; $p = 0.179$).

All the cases were sporadic VSs except for six patients who were affected by neurofibromatosis type 2 (NF2). Seventy-two VSs were diagnosed on the right side and 88 VSs on the left side. Each patient received an MRI scan not exceeding 1 month before admission. Tumors were measured in three spatial dimensions (on axial and coronal MRI section planes), and tumor size was estimated considering the major diameter, including the part of tumor extending into the internal auditory canal (IAC); mean tumor size was 23.3 mm. Twenty-three (14.4%) cystic tumors

L. Mastronardi (✉) · A. Campione · G. Cacciotti · R. Roperto · C. Giacobbo Scavo
Department of Neurosurgery, San Filippo Neri Hospital—ASLRoma1, Rome, Italy
e-mail: mastro@tin.it

© Springer Nature Switzerland AG 2019
L. Mastronardi et al. (eds.), *Advances in Vestibular Schwannoma Microneurosurgery*, https://doi.org/10.1007/978-3-030-03167-1_8

were reported, and their mean size (30.9 mm) was found to be higher than that of solid tumors (22.1 mm) with a high statistical significance ($p = 0.0003$) (Table 8.1).

Facial nerve (N VII) function was evaluated preoperatively according to the House-Brackmann (HB) grading system [1]. One hundred thirty-five patients (84.4%) had completely preserved N VII function (HB 1) before surgery; 14 patients had slight functional impairment (HB 2), while two were deemed HB 3 and three were deemed HB 4 (data of the remaining six patients not available).

Audiological exams were performed the day before surgery and 1 week and 6 months after by pure tone audiometry (PTA), auditory brainstem response (ABR), and monosyllabic speech audiograms. The results were interpreted according to the American Academy of Otolaryngology-Head and Neck Surgery (AAO-HNS) hearing classification. AAO-HNS Hearing Classification is based on the evaluation of both PTA and speech discrimination score (SDS); class A corresponds to PTA ≤ 30 dB and SDS $\geq 70\%$ and class B to PTA ≤ 50 dB and SDS $\geq 50\%$. The same cutoff values are used to define I (good-excellent hearing) and II (serviceable hearing) grade of the Gardner-Robertson Scale [2], respectively; therefore, when "serviceable" hearing (SH) is mentioned, AAO-HNS classes A and B are intended. At the preoperative evaluation, 8 patients were class A, 48 patients were class B, 68 patients were class C, and 33 patients were class D (data of the remaining six patients not available). Therefore, 56 (35%) patients had preoperative serviceable hearing. Tumor size at diagnosis significantly correlated with preoperative hearing function retention, as already reported in the literature [3]: patients AAO-HNS class A had significantly smaller tumors than those AAO-HNS class B ($p = 0.005$), and these latter had significantly smaller tumors than those AAO-HNS classes C ($p = 0.0008$) and D ($p = 0.002$) (Table 8.2).

Table 8.1 Mean tumor sized of the overall series and as stratified according to tumor consistency (solid or cystic)

	Number of tumors	Mean tumor size (mm)	Statistical significance
Overall series	160	23.3	–
Solid tumors	137	22.1	
Cystic tumors	23	30.9	vs solid tumors, $p = 0.0003$

Table 8.2 Correlation between preoperative hearing status and mean tumor size at diagnosis

	Number of patients	Mean tumor size (mm)	Statistical significance
AAO-HNS class A	8	11.1	–
AAO-HNS class B	48	19.6	vs mean tumor size of class A, $p = 0.005$
AAO-HNS class C	68	25.3	vs mean tumor size of class B, $p = 0.0008$
AAO-HNS class D	33	25.9	vs mean tumor size of class B, $p = 0.002$

8.2   Intraoperative Data

Retrosigmoid approach was performed in all the cases except for one, wherein the patient underwent a translabyrinthine approach. The extent of tumor removal was evaluated intraoperatively and by postoperative Gd-enhanced MRI 24–48 h after surgery. Total resection was reported in 64 cases, near-total resection (99%: thin capsule residue left on the brain stem) in 33, subtotal resection (extent of removal between 90 and 99%) in 46, and partial resection (extent of removal <90%) in 17. Thus, the total or near-total resection rate was 60.6%. The mean operating time was approximately 5 h and the mean blood loss was <200 mL.

The anatomy of the facial nerve was carefully studied intraoperatively, and the course of the nerve could be traced in 147 cases with the aid of the bipolar stimulator; the anterosuperior course pattern was the most common (40.8%), followed by the anterior (34.7%), the anteroinferior (23.8%), and the dorsal pattern, which was observed in one case only [4]. This data is in line with a previous report by our team [5]. On the contrary, Sameshima et al. [6] described a series wherein the anterior pattern was the most common (52% of cases), followed by anterosuperior (38.5%) and anteroinferior (5.3%).

The N VII was anatomically preserved and responded to functional stimulation in 150 (93.7%) cases (comprising the only case of translabyrinthine approach). Four cases of anatomical preservation and functional blockade were reported. In five cases, the N VII was interrupted during surgery (data of the remaining patient not available).

The cochlear nerve was anatomically preserved and responded to ABR stimulation in 64 (40%) cases; anatomical preservation with transient or permanent functional blockade was reported in 18 cases. Among the 56 cases where SH was retained preoperatively, 40 (71.4%) cases of both anatomical and functional preservation were reported (Table 8.3).

In 14 surgical procedures, a flexible endoscope was employed. At the end of classical microsurgical resection, a 4-mm flexible video endoscope (105 4 mm × 65 cm, Karl Storz, GmbH, Tuttlingen, Germany) was inserted in the surgical cavity, handled by the operator. The endoscope was introduced under both microscopic and endoscopic visualization to prevent injury to cerebellopontine angle (CPA) structures, and the endoscopic tip was oriented into the IAC in order to detect tumor residue hiding in the deeper portion of IAC. If residual tumor was identified,

Table 8.3 Intraoperative anatomical and functional preservation of the cranial nerves

	Anatomical and functional preservation	Only anatomical preservation	No preservation feasible
Facial nerve (data of 1 patient not available)	149/160 (93.1%)	5/160 (3.1%)	5/160 (3.1%)
Cochlear nerve (overall series)	64/160 (40%)	18/160 (11.3%)	79/160 (49.8%)
Cochlear nerve (preoperative SH patients[a])	40/56 (71.4%)	8/56 (14.3%)	8/56 (14.3%)

[a]SH patients, i.e., patients with preoperative AAO-HNS class A-B of hearing

microsurgical resection was pursued, and further endoscopic controls were repeated until complete tumor resection was accomplished [7, 8]. In all (100%) of the flexible endoscope-assisted procedures, total or near-total resection was reported, and the postoperative MRI scans confirmed that no residual traces of tumor were present within the IAC [7, 8].

In 67 surgical procedures, a laser device was used. In particular, the capsule incision and tumor debulking were performed with handheld 2μ thulium flexible laser fiber (Revolix jr; Lisa laser). The range of power setting was 1–14 W. Standard 0.9% saline solution irrigation was used to cool the fiber. The fiber was used for cutting, vaporizing, and coagulating the capsule and the intracapsular mass, in combination with bipolar forceps, microscissors, and Sonopet Ultrasonic Aspirator. Following tumor debulking, the remaining tumor capsule is removed with standard microsurgical tools [9]. Forty-three (64.3%) cases of total or near-total resection were reported; although such a resection rate was higher than that observed in not-laser-assisted procedures (58.1%), the difference was not statistically significant ($p = 0.435$).

8.3   Postoperative Data

Facial nerve function was assessed both clinically and with electromyography (EMG) 1 week and 6 months postoperatively. Among the 159 patients who underwent a retrosigmoid craniotomy, 75 (47.2%) had HB 1 facial nerve function at 1 week after surgery; 74 patients had only minor functional impairment (HB 2–3) that completely resolved in the following 6 months after surgery. At this time, 149 (93.7%) patients had intact or completely recovered N VII function—a preservation rate in accordance with previous literature [10–16]. As stratified according to the course pattern of the facial nerve, HB 1 facial nerve function was observed in 95% of VSs with a N VII running anterosuperiorly or anteroinferiorly; on the contrary, facial functional prognosis was slightly worse in case of purely anterior course pattern, with 84% of HB 1 facial nerve function (Table 8.4).

Audiological exams were performed 1 week and 6 months postoperatively by PTA, ABR, and monosyllabic speech audiograms. Of the 56 patients who had SH before surgery, 35 retained SH after surgery, thus yielding a hearing preservation (HP) rate of 62.5%, which is in line with diverse reports from the literature [11–19].

Table 8.4 Correlation between facial nerve course patterns and functional preservation in the long term

	Anterosuperior (%)	Anterior (%)	Anteroinferior (%)	Dorsal (%)
Course pattern frequency	40.8	34.7	23.8	0.7[a]
HB 1 facial nerve function rate at 6 months	95	84	95	0[a]

[a]The features refer to a single patient who had HB 3 facial nerve function at 6 months, although with signs of improvements

Mean tumor size at diagnosis of those cases where HP could be obtained was 18 mm, significantly lower ($p = 0.0002$) than that of tumors where HP was not feasible during surgery or impossible due to preoperative hearing deterioration. Thus, postoperative HP rate in our series correlated with both preoperative hearing status and tumor size at presentation.

In 63 procedures, diluted papaverine was used. Papaverine at a concentration of 0.3% was administered topically after tumor debulking and in case of nerve traction to allow for functional recovery and microvascular protection of the cranial nerves. The functional outcome concerning postoperative N VII function was in line or even superior to that observed in the overall series: at 1 week after surgery, 30 (43.5%) patients had HB 1 facial nerve function, which raised to 61 (96.8%) at 6 months of follow-up. As far as HP is concerned, 19 patients had SH before surgical intervention; among these latter, 10 (52.6%) received successful HP. Although such a HP rate is lower than that observed both in the general series and in the procedures wherein papaverine was not employed, this difference did not reach statistical significance ($p = 0.47$) after a chi-square test. However, we do not have an explanation for this phenomenon, and further studies are warranted to clarify the role of papaverine in VS surgery.

As regards the 67 laser-assisted procedures, functional outcomes were ultimately in line with those observed in the whole series. Twenty-six (38.8%) patients had HB 1 facial nerve function 1 week after surgery; the temporary worse result of immediately postoperative facial function seemed to be in relation to size of VSs, as already reported in a previous article of ours [9]. In fact, in patients with immediate normal face (HB 1), mean tumor size was 18 mm versus 28 mm of cases with transient HB 2 to HB 4 facial palsy recovered within 6 months ($p = 0.0000008$). At 6 months, indeed, facial nerve preservation rose to 65 (97%) patients. HP was attempted in 18 (26.9%) patients, 12 (66.6%) of whom eventually retained SH after surgery (Table 8.5).

Postoperative complications were sporadically observed, the most common being cerebrospinal fluid (CSF) leakage that affected 17 (10.7%) of the 159 patients who underwent a retrosigmoid craniotomy and mostly presented as rhinoliquorrhea, in ten cases—in line with other reports [20–22]. Eight (5%) cases of wound infection were reported, four of them requiring revision surgery. Rarer complications were cerebellar mutism (two cases), prolonged vertigo (seven cases), diplopia for

Table 8.5 Functional outcomes in the overall series and in subgroups employing adjunct devices/drugs

	Overall series	Papaverine group	Laser group
HB 1 facial nerve function, 1 week postoperatively	75/159 (47.2%)	30/63 (43.5%)	26/67 (38.8%)
HB 1 facial nerve function, 6 months postoperatively	149/159 (93.7%)	61/63 (96.8%)	65/67 (97%)
AAO-HNS class A-B of hearing preservation rate	35/56 (62.5%)	10/19 (52.6%)	12/18 (66.6%)

abducens paresis (five cases), pneumonia (one case), and hydrocephalus (one case); all the neurological complications were transient, and the only case of hydrocephalus was promptly solved with external CSF drainage.

HAC fluid cement was used in 14 surgical procedures to fill the void around the bone flap at the end of reconstructive cranioplasty, which consisted of repositioning the bone flap and fixing with titanium screws. As reconstructive retrosigmoid cranioplasty was combined with an underlay hourglass-shaped autologous pericranium duraplasty—as already reported in a recent study of ours [23]—no postoperative wound infections and meningitis were observed; only one (7.1%) case of CSF fistula was reported.

References

1. House JW, Brackmann DE. Facial nerve grading system. Otolaryngol Head Neck Surg. 1985;93(2):146–7.
2. Gardner G, Robertson JH. Hearing preservation in unilateral acoustic neuroma surgery. Ann Otol Rhinol Laryngol. 1988;97(1):55–66.
3. Hoa M, Drazin D, Hanna G, Schwartz MS, Lekovic GP. The approach to the patient with incidentally diagnosed vestibular schwannoma. Neurosurg Focus. 2012;33(3):E2.
4. Nejo T, Kohno M, Nagata O, Sora S, Sato H. Dorsal displacement of the facial nerve in acoustic neuroma surgery: clinical features and surgical outcomes of 21 consecutive dorsal pattern cases. Neurosurg Rev. 2016;39(2):277–88; discussion 88.
5. Mastronardi L, Cacciotti G, Roperto R, Di Scipio E, Tonelli MP, Carpineta E. Position and course of facial nerve and postoperative facial nerve results in vestibular schwannoma microsurgery. World Neurosurg. 2016;94:174–80.
6. Sameshima T, Morita A, Tanikawa R, Fukushima T, Friedman AH, Zenga F, et al. Evaluation of variation in the course of the facial nerve, nerve adhesion to tumors, and postoperative facial palsy in acoustic neuroma. J Neurol Surg B Skull Base. 2013;74(1):39–43.
7. Corrivetti F, Cacciotti G, Scavo CG, Roperto R, Mastronardi L. Flexible endoscopic-assisted microsurgical radical resection of intracanalicular vestibular schwannomas by retrosigmoid approach: operative technique. World Neurosurg. 2018;115:229–33.
8. Mastronardi L, Cacciotti G, Scipio ED, Parziale G, Roperto R, Tonelli MP, et al. Safety and usefulness of flexible hand-held laser fibers in microsurgical removal of acoustic neuromas (vestibular schwannomas). Clin Neurol Neurosurg. 2016;145:35–40.
9. Mastronardi L, Cacciotti G, Roperto R, Tonelli MP, Carpineta E, How I. Do it: the role of flexible hand-held 2μ-thulium laser fiber in microsurgical removal of acoustic neuromas. J Neurol Surg B Skull Base. 2017;78(4):301–7.
10. Cardoso AC, Fernandes YB, Ramina R, Borges G. Acoustic neuroma (vestibular schwannoma): surgical results on 240 patients operated on dorsal decubitus position. Arq Neuropsiquiatr. 2007;65(3A):605–9.
11. Roessler K, Krawagna M, Bischoff B, Rampp S, Ganslandt O, Iro H, et al. Improved postoperative facial nerve and hearing function in retrosigmoid vestibular schwannoma surgery significantly associated with semisitting position. World Neurosurg. 2016;87:290–7.
12. Samii M, Gerganov V, Samii A. Improved preservation of hearing and facial nerve function in vestibular schwannoma surgery via the retrosigmoid approach in a series of 200 patients. J Neurosurg. 2006;105(4):527–35.
13. Samii M, Matthies C. Management of 1000 vestibular schwannomas (acoustic neuromas): the facial nerve–preservation and restitution of function. Neurosurgery. 1997;40(4):684–94; discussion 94–5.

14. Tatagiba MS, Roser F, Hirt B, Ebner FH. The retrosigmoid endoscopic approach for cerebellopontine-angle tumors and microvascular decompression. World Neurosurg. 2014;82(6 Suppl):S171–6.
15. Tatagiba M, Roser F, Schuhmann MU, Ebner FH. Vestibular schwannoma surgery via the retrosigmoid transmeatal approach. Acta Neurochir. 2014;156(2):421–5; discussion 5.
16. Yang J, Grayeli AB, Barylyak R, Elgarem H. Functional outcome of retrosigmoid approach in vestibular schwannoma surgery. Acta Otolaryngol. 2008;128(8):881–6.
17. Ahsan SF, Huq F, Seidman M, Taylor A. Long-term hearing preservation after resection of vestibular schwannoma: a systematic review and meta-analysis. Otol Neurotol. 2017;38(10):1505–11.
18. Mazzoni A, Zanoletti E, Calabrese V. Hearing preservation surgery in acoustic neuroma: long-term results. Acta Otorhinolaryngol Ital. 2012;32(2):98–102.
19. Nakamizo A, Mori M, Inoue D, Amano T, Mizoguchi M, Yoshimoto K, et al. Long-term hearing outcome after retrosigmoid removal of vestibular schwannoma. Neurol Med Chir (Tokyo). 2013;53(10):688–94.
20. Ansari SF, Terry C, Cohen-Gadol AA. Surgery for vestibular schwannomas: a systematic review of complications by approach. Neurosurg Focus. 2012;33(3):E14.
21. Bennett M, Haynes DS. Surgical approaches and complications in the removal of vestibular schwannomas. Otolaryngol Clin N Am. 2007;40(3):589–609, ix–x.
22. Nonaka Y, Fukushima T, Watanabe K, Friedman AH, Sampson JH, Mcelveen JT, et al. Contemporary surgical management of vestibular schwannomas: analysis of complications and lessons learned over the past decade. Neurosurgery. 2013;72(2 Suppl Operative):ons103–15; discussion ons15.
23. Mastronardi L, Cacciotti G, Caputi F, Roperto R, Tonelli MP, Carpineta E, et al. Underlay hourglass-shaped autologous pericranium duraplasty in "key-hole" retrosigmoid approach surgery: technical report. Surg Neurol Int. 2016;7:25.

Part III

New Technologies

Intraoperative Identification and Location of Facial Nerve: Type of Facial Nerve Displacement—How to Use Monopolar Stimulator

9

Luciano Mastronardi, Alberto Campione, Ali Zomorodi,
Ettore Di Scipio, Antonio Adornetti,
and Takanori Fukushima

9.1 Intraoperative Identification and Location of Facial Nerve: Position, Course, and Functional Preservation

Intraoperative facial nerve monitoring (IOFNM) is a neurophysiological method whose main purpose is to inform the surgical team of the actual neural function of the facial nerve (N VII) so that the operative strategy can be consequently adjusted to avoid neural damage. Functional changes in the activity of N VII also have a role in assessing postoperative functional prognosis.

As electrical stimuli are provided locally in order to elicit a functional response, IOFNM can be used to identify N VII and trace its course along the surgical field, thus letting the surgeon reduce undue nerve injury.

The most commonly used—and therefore defined as "standard"—IOFNM techniques are direct electrical stimulation (DES) and free-running electromyography (EMG).

L. Mastronardi (✉) · A. Campione
Department of Neurosurgery, San Filippo Neri Hospital—ASLRoma1, Rome, Italy
e-mail: mastro@tin.it

A. Zomorodi · T. Fukushima
Division of Neurosurgery, Duke University Medical Center, Carolina Neuroscience Institute, Raleigh, NC, USA
e-mail: ali.zomorodi@duke.edu; Fukushima@carolinaneuroscience.com

E. Di Scipio
Department of Neurology and Neurophysiology, San Filippo Neri Hospital—ASLRoma1, Rome, Italy

A. Adornetti
Department of Neurosurgery, San Filippo Neri Hospital - ASLRoma1, Roma, Italy
e-mail: adornetti@hospitaldevice.it

© Springer Nature Switzerland AG 2019

L. Mastronardi et al. (eds.), *Advances in Vestibular Schwannoma Microneurosurgery*, https://doi.org/10.1007/978-3-030-03167-1_9

9.1.1   Guidelines on Intraoperative Facial Nerve Monitoring

The most recent evidence-based guidelines by the Congress of Neurological Surgeons recommend that IOFNM be routinely used during VS surgery to improve long-term facial function [1]. However, no specific recommendation has been proposed about preferable neurophysiological techniques; in the absence of any clinical trial directly comparing the three techniques discussed above, a combination of them seems a wise choice and has already been successfully experimented in many studies [2–5].

As far as functional prognosis is concerned, the guidelines state that IOFNM can be used to accurately predict favorable long-term N VII function. Specifically, the presence of favorable testing reliably portends a good long-term facial nerve outcome. However, the absence of favorable testing in the setting of an anatomically intact N VII does not reliably predict poor long-term function and therefore cannot be used to direct decision-making regarding the need for early reinnervation procedures [1]. In spite of numerous studies reporting valid predictive factors for long-term facial outcome [2–4], the lack of monitoring standardization and the common empiric observation of functional improvement in patients with early facial palsy do not allow for conclusive decisions about the predictive value of unfavorable IOFNM results. Therefore, poor intraoperative EMG electrical response of N VII should not either be used as a reliable predictor of poor long-term facial function.

9.2   Standard Techniques: Essential Technical Considerations [6]

Standard IOFNM techniques both operate on the same general principle and share the same technical apparatus: an electrical stimulus is either given through a probe or spontaneously conducted by N VII, and the response is recorded by recording electrodes and/or motion detectors.

The type of electrical stimulus employed in IOFNM is a rectangular pulse that administers a certain intensity of current (measured in milliampere, mA) during a definite interval of time (measured in milliseconds, ms), thus delivering an amount of charge (measured in Coulomb, C) that is equal to the product of pulse intensity and pulse duration. Two types of stimulators are available, and the main difference is whether they maintain a constant current or a constant voltage during stimulation. The debate on which one should be preferred is still controversial and inconclusive.

Stimulation probes directly come into contact with targets to elicit a response. Two electrodes are always needed to produce current; however, two different configurations of stimulation exist. In monopolar stimulation, an active electrode contacts the target while a reference electrode is placed far apart; instead, in bipolar stimulation, both electrodes are active and contact the target so that the current flows through it. The current produced by the two different probes has peculiar characteristics in terms of intensity needed to reach a response, density distribution, and local spreading/shunting. Monopolar probes usually require a current intensity two to three times higher than bipolar probes. The current density is defined as the amount of electrical current per unit area of cross section; its distribution represents the diverse values that current intensity reaches in the stimulated area. Monopolar probes are characterized by a better and more predictable current density distribution, and this leads to a direct correlation between stimulus strength and response intensity. However, as electrical current spreads across the target tissue, it may activate any excitable tissue other than the target, thus leading to false-positive responses which are the main disadvantage of monopolar electrodes. On the contrary, with bipolar probes, the target is in between the two electrodes, and current spreading is minimal, thus leading to a more precise and localized stimulation which is the main advantage of this type of probes. The main disadvantage of bipolar stimulation occurs when a less resistive substance interposes between the electrodes and causes current shunting. A practical example may be the case of surgical field irrigation or temporary bleeding: the current flows mainly through the liquids and this leads to a false-negative response. In general, monopolar probes provide responses with high sensitivity, and bipolar probes are more suitable when a high degree of specificity is required.

Electroencephalography (EEG) platinum needle electrodes are the most commonly used recording electrodes in IOFNM because of their ability to detect muscular activity anywhere in the target muscle. Multichannel electrode montage has become the standard EMG recording configuration as it increases the sensitivity of the monitoring; it consists of positioning a pair of bipolar recording electrodes (i.e., a recording channel) in both the orbicularis oculi and the orbicularis oris muscles to monitor the superior and inferior N VII branches, respectively. For the orbicularis oculi, the electrodes should be inserted at the lateral canthus, below the eyebrow, and at a distance of 1.5 cm from the orbital rim. For the orbicularis oris, the first electrode is inserted at a distance of 2 cm from the oral commissure; the second electrode is positioned 1 cm apart within either the superior or the inferior lip.

A major limitation of EMG recording techniques is the large number of artifacts caused by electrocautery that compromise the identification of eventual N VII thermal injury. Motion detecting devices have been employed to identify abnormal nerve activity on the basis on facial muscle contractions during electrocautery, but none of them allow adequate monitoring at the state of the art. Therefore, this surgical step still remains a threat to N VII functional preservation.

Facial muscle activity is fundamental to the effectiveness of EMG recordings; therefore, anesthetic regimens based on neuromuscular blocking agents may interfere with the potential propagation and, in the end, with the monitoring itself. However, the use of short-acting agents during tracheal intubation is generally believed to be secure as the drug is expected to be cleared during the surgical approach and well before the surgical steps that are critical to IOFNM.

9.2.1 Direct Electrical Stimulation (DES)

DES [6] consists of applying an electrical stimulus to N VII by using a probe in order to obtain a triggered EMG-recorded response. The electrical stimulation generates compound muscle action potentials (CMAPs) that are recorded by electrodes positioned within the orbicularis oculi and the orbicularis oris muscles; a third reference electrode is positioned at the forehead. The response of the facial muscles is monitored acoustically through a loudspeaker, and its wave parameters—latency and amplitude—are observed on the monitor. The principle at the basis of DES is that higher current intensity is needed to obtain a response when N VII is injured or when the nerve is stimulated at the root entry zone (REZ). Thus, not only does DES allow N VII localization and course tracing, but it also provides qualitative evaluation of N VII anatomic and functional integrity.

9.2.1.1 DES: Localization of Facial Nerve

Localization of N VII in the cerebellopontine angle (CPA) does not follow standardized protocols, yet it is always wise to start with stimuli in the range of 1–3 mA—for a first and gross screening—and progressively reduce the stimulation intensity while approaching the nerve. Indeed, when N VII is exposed, even a very mild stimulation in the range of 0.1–0.2 mA guarantees precise localization and prevention of both electrical/thermal injury and current spreading. The latter is a major disadvantage of monopolar probes, and its entity is not insignificant: it is calculated that a monopolar 1 mA stimulus could spread across 1 mm of temporal bone covering N VII as well as a monopolar 0.5–0.6 mA stimulus could spread around at 2 cm of distance, possibly eliciting a false-positive response. Once N VII has been localized, DES should be performed periodically during surgery so as to confirm the course and the functional activity of the nerve.

In spite of its lack of target specificity, current spreading due to monopolar stimulation can be turned to the surgeon's advantage when used on the tumor capsule: if no response is reached, it is then excluded that the nerve courses superficially under the capsule or as splayed fibers on the capsule itself (a procedure known as "facial nerve detector technique"). Indeed, this procedure allows to spare N VIIs with distorted courses and is especially valuable in the case of a dorsally displaced nerve. Dorsal N VII displacement is very rare but also very problematic because the tumor has to be manipulated beside/behind the nerve and facial outcome may be therefore compromised by surgical trauma. As Nejo et al. [7] and Sameshima et al. [8] reported, dorsal course pattern is very rare (3.8% and 0.3% of cases, respectively) and characterizes N VIIs in the presence of middle-large VSs (28 mm of mean diameter and 1.5–3 cm range, respectively). Although no statistically significant differences were reported when dorsal pattern was compared to more common nerve courses in terms of long-term postoperative facial function, other important features affected the outcomes of VS surgery in such cases. The rate of total or near-total resection in the group of VSs with dorsal N VII displacement (D group) was significantly inferior that in the non-dorsal displacement group (ND group): 38% vs. 85.4%, ($p < 0.0001$); conversely, the retreatment rate was significantly higher in the D group than in the ND group: 33.3% vs. 1.3% ($p < 0.0001$) [7]. As Nejo et al. further analyzed the D group, N VII morphology emerged as a prominent factor influencing both tumor resection extent and retreatment rate: the cases of VS with dorsally displaced and broadened nerve were significantly more difficult to resect totally or near-totally and had to undergo a reoperation more frequently than the cases wherein the nerve was morphologically intact [7]. Although only one dorsally displaced pattern was observed in Sameshima et al., it showed, along with rostral and caudal displacement patterns, significantly stronger adhesion of N VII to the tumor capsule, which would eventually negatively affect the facial functional outcome [8].

Complete data about the diverse course patterns of N VII were reported by Sameshima et al. [8], who classified the nerve paths into six categories: ventral central, ventral rostral, ventral caudal, rostral, caudal, and dorsal. They reported that ventral central pattern was the most common (52% of cases), followed by ventral rostral (38.5%) and ventral caudal (5.3%). Despite the relative frequency of these categories remained constant even after stratification into classes of tumor diameter, the ventral rostral pattern was more common in larger tumors—a sign, according to the authors, of progressive nerve displacement due to tumor growth. Rostral, caudal, and dorsal course patterns were rare and significantly associated to stronger adhesion of N VII to tumor capsule. Interestingly, no significant differences in terms of late postoperative facial function were reported among the six path categories. Indeed, the nerve path did not show as a predictive factor of facial paresis per se but only when associated to capsule adhesiveness.

In previous articles [5, 9], we classified the N VII course patterns into four categories, anterosuperior (AS), anterior (A), anteroinferior (AI), and dorsal (D), and found a different pattern distribution in a 100-patient series: AS course was most common (48%), followed by A (31%) and AI (21%) patterns (Fig. 9.1).

No dorsal course pattern cases were observed in this series. It was also found that distorted course patterns such as A and AI were more common in larger tumors and supposedly due to tumor enlargement itself. In fact, as VSs most commonly arise from the inferior vestibular nerve, an anterosuperior displacement of N VII would be logically expected, and tumor growth may have the effect of pushing the nerve in other directions. Tumors with A nerve course also showed greatest growth tendency and significantly inferior late postoperative facial function recovery rate as compared with the two alternative patterns.

9.2.1.2 DES: Functional Preservation of Facial Nerve and Functional Outcome Predictive Values

The electrical stimulation of N VII generates compound muscle action potentials (CMAPs) that are recorded as EMG waves characterized by their own latency and amplitude. The CMAP amplitude is directly proportional to the number of N VII fibers that have been stimulated and are therefore viable, intact; thus, if amplitude decreases, the functional preservation of N VII is at risk [6]. Amano et al. studied the correlation between CMAP amplitude and early postoperative facial function. The maximal response after stimulation at REZ was recorded and measured in

Fig. 9.1 Position and
course of N VII and its
relationship with VS. *AS*
anterosuperior, *D* dorsal, *A*
anterior, *AI* anteroinferior.
*Reprinted from World
Neurosurgery, 94, Luciano
Mastronardi, Guglielmo
Cacciotti, Raffaelino
Roperto, Ettore di Scipio,
Maria Pia Tonelli, Ettore
Carpineta, Position and
Course of Facial Nerve
and Postoperative Facial
Nerve Results in Vestibular
Schwannoma
Microsurgery, Pages No.
174–180, 2016, with
permission from Elsevier*

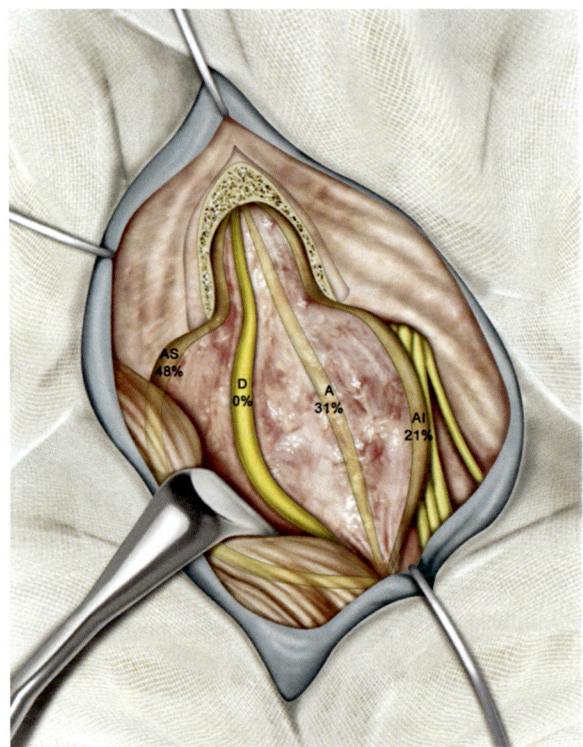

microvolts (μV) after tumor removal. It was observed that the maximal response for muscles with slight palsy after surgery was around 1000 μV; maximal responses above that value were correlated with no cases of early facial palsy, and therefore the author proposed 1000 μV as the cutoff under which tumor removal posed N VII functional preservation at risk ("warning criterion"). The tumor removal would eventually be suspended if maximal response reached a value <800 μV ("tumor resection limit") [2]. Duarte-Costa et al. [3] concentrated on the correlation between CMAP amplitude and long-term facial function in cases of Grade IV VSs. A statistically significant difference was observed between the good and the bad prognosis groups in terms of mean "proximal amplitude"—i.e., amplitude of CMAP elicited at REZ—and a cutoff of 420 μV was proposed. Higher values of amplitude would predict a good long-term facial function (House-Brackmann grades 1–2; Table 9.1) with a sensitivity of 73% and a specificity of 67%; on the contrary, lower values would have a negative predictive value of 79%.

CMAPs are responses to stimuli that can be provided at different sites, most importantly at REZ or at the internal acoustic canal (IAC); the amplitude of CMAPs elicited at REZ decreases during surgery, while those evoked at IAC remain almost constant. As absolute amplitude values are interindividually variable, calculating the proximal/distal (i.e., REZ/IAC) amplitude ratio after tumor removal is a means to solve this problem and provide more standardized cutoffs to adjust the surgical

Table 9.1 House-Brackmann (HB) facial function grading system

Grade	Description	Gross function	At rest	In motion
1	Normal	Normal	Normal	Normal
2	Mild dysfunction	Slight weakness with effort; may have slight synkinesis	Normal	Mild oral and forehead asymmetry; complete eye closure with minimal effort
3	Moderate dysfunction	Obvious asymmetry with movement; noticeable synkinesis or contracture	Normal	Mild oral asymmetry; complete eye closure with effort; slight forehead movement
4	Moderately severe dysfunction	Obvious asymmetry; disfiguring asymmetry	Normal	Asymmetric mouth; incomplete eye closure; no forehead movement
5	Severe dysfunction	Barely perceptible movement	Asymmetric	Slight oral/nasal movement with effort; incomplete eye closure
6	Total paralysis	None	Asymmetric	No movement

House JW, Brackmann DE. Facial nerve grading system. Otolaryngol Head Neck Surg. 1985;93(2):146–7

strategy in such situations wherein N VII preservation may be jeopardized. In their systematic review, Acioly et al. [6] reported different predictive values: ratio >30% would predict good long-term facial function, but a more complex risk stratification is also available. According to the latter, ratio >90% predicts good short- and long-term facial function; ratio 50–90% predicts poor short-term functional outcome with long-term recovery; finally, ratio <50% predicts even long-term poor results. These figures are in line with the study by Duarte-Costa et al. [3]: after finding a significant difference in terms of ratio between good and bad prognosis groups, a cutoff predictive of poor long-term facial results for Grade IV VSs was proposed. A ratio <44% would predict long-term facial palsy with positive predictive value of 85%, sensitivity of 73%, and specificity of 78%. Amano et al. [2] calculated a different type of ratio; in their study, a ball-type monopolar electrode was used to elicit maximal responses before surgery (control maximal amplitude) and after tumor removal (last maximal amplitude). The amplitude preservation ratio was defined as last maximal amplitude/control maximal amplitude and was correlated to the risk of long-term facial palsy. With a ratio >50%, less than 5% of patients suffered from poor late facial function; therefore, this value was proposed as a cutoff under which N VII functional preservation would be at risk (warning criterion). On the contrary, the risk of long-term facial function impairment dramatically rose to almost 25% for ratio <40%; this was proposed as a tumor resection limit cutoff.

A different approach to N VII preservation is not to measure how many fibers are still intact after the surgical procedure but how well the remaining fibers conduct impulses. Seeking the smallest amount of current necessary to elicit an EMG-recordable response—i.e., finding the threshold—allows for a semiquantitative functional evaluation of the nerve activity: the lower the threshold is, the higher is the conductance (and thus the viability) of the nerve. Low thresholds are <0.05–0–1 mA, while high thresholds reach 2–3 mA and are predictive of poor outcome

[6]. It has also been reported [10] that strong nerve adhesion correlates with low mean stimulation threshold: therefore, tumors difficult to dissect also compromise N VII function, supposedly because stretched fibers are more prone to neurotmesis. Although measuring the stimulation threshold offers an interesting insight into qualitative profile of N VII activity, it is not a standardized procedure yet, and different studies use different stimulation protocols to find the threshold itself.

9.2.1.3 Limits of DES

Triggered CMAP obtained with DES can only be used intermittently, and stimulation at REZ—fundamental to the calculation of proximal/distal amplitude ratio—can only be performed after the identification of the nerve at the brainstem. This is particularly difficult in patients with large tumors, because of anatomical distortion of the brainstem and late identification of the proximal N VII that is inaccessible during most of the surgical procedure. Therefore, proximal N VII identification and the recording of the amplitude ratio cannot be performed in 30–35% of monitored patients owing to technical reasons, distorted anatomy, or surgical approach [6].

9.2.2   Free-Running EMG

Continuous free-running EMG [6] consists of recording facial muscle activity responses that are elicited by either surgical maneuvers or spontaneous N VII discharge. EMG activity is also monitored acoustically through loudspeakers so that characteristic patterns are also recognizable according to their acoustic quality. The most important type of EMG recording during surgery is the neurotonic discharge, which comprises muscle activity in response to mechanical or metabolic irritation of N VII. Interestingly, similarly to triggered CMAPs, injured motor nerves are less likely to evoke neurotonic discharges after mechanical trauma.

The patterns of EMG activity are classified into spontaneous and evoked activity. The evoked activities correspond to EMG responses that occur as a direct consequence of surgical maneuvers such as DES, mechanical trauma, and electrocautery. EMG activities of different amplitudes and waveforms are further subdivided into bursts, trains, and pulse EMG patterns [6].

The pulse pattern is observed after DES and is characterized by pulsed sounds synchronous with the electrical stimulation. The burst pattern is the most frequently encountered EMG activity and consists of short, relatively synchronous bursts of motor unit potentials that last up to a few 100 ms. The burst pattern arises as a consequence of maneuvers like DES, electrocautery, or irrigation and is probably due to the mechanoreceptor properties or metabolic irritation of the N VII fibers resulting in depolarization and elicitation of the action potential. Therefore, burst activity is an indirect sign of a still functional N VII as severely injured fibers would not be able to conduct electrical impulses. Finally, the train pattern is characterized by asynchronous trains of motor unit potentials with a duration of up to several minutes. Two types of trains have been identified: high-frequency trains (50–100 Hz), which have a typical acoustic quality that resembles an airplane engine; and low-frequency trains (1–50 Hz), which are associated with acoustic signals similar to popping popcorn and are rarer than high-frequency trains. Train activity is mostly correlated to the surgical traction of N VII, especially when the traction occurs in a lateral-to-medial direction within the CPA. The train activity arises from seconds to minutes after provoking events like electrocautery, mild nerve trauma, and free irrigation. Train responses are frequently observed in cases of N VII strong adhesion or encasement; the nerve then becomes more susceptible to dissection and surgical traction. However, the onset delay does not always allow to establish a direct cause-and-effect relationship between surgical manipulation and EMG responses, thus making real-time surgical strategy modification unfeasible [6].

Romstöck et al. [11] introduced a more complex classification of spontaneous EMG activity patterns, focusing especially on different types of train activity. According to their terminology, spikes correspond to biphasic or triphasic potentials with one large peak (of amplitude). Bursts are defined as an isolated complex of superimposed spikes arranged in a spindle-like fashion that shows several large peaks of up to 5000 μV and lasts up to several 100 ms. A-train is a unique sinusoidal waveform pattern with a typical high-frequency acoustic signal that always has a sudden onset, amplitude never exceeding 500 μV, a frequency of 60–200 Hz, and duration of milliseconds to several seconds. B-train is a regular or irregular sequence of a single spike or burst component that has a gradual onset and a long duration, from 500 ms to hours. C-train is characterized by a continuous EMG irregular activity. Spikes and bursts occur immediately after direct mechanical trauma from surgical instruments near N VII and together with B- and C-trains are clinically irrelevant. On the contrary, the occurrence of A-trains is associated with N VII injury. The A-train EMG pattern is highly suggestive of repetitive discharges that are found in chronic denervation processes and myopathies. Therefore, after nerve injury, the corresponding muscle fibers may become unstable and serve as a pacemaker because they are no longer under neural. The first occurrence of A-trains can always be correlated with specific surgical maneuvers, especially the dissection of the tumor surface near the brainstem and IAC decompression [6, 11].

The main advantage of intraoperative free-running EMG is that it provides the surgeon with almost real-time feedback of any surgical maneuver that could result in nerve injury. Indirectly, spontaneous EMG activity can even help localize N VII by warning the surgeon about nerve proximity even when it has not been exposed yet. These considerations are valid for spikes (or pulses) and bursts, which are synchronous to mechanical trauma and do not show any relevant delay; however, they are not directly correlated to nerve injury, and the only pattern that shows this correlation, i.e., A-train activity, arises with considerable delay. Thus, A-train activity may not be a reliable parameter for nerve preservation, but it has been investigated as a predictive factor of facial function prognosis although the lack of standardization of EMG patterns in the literature does not allow for definitive conclusions about its role. Especially Romstöck et al. [11] identified A-train activity in almost all of the patients affected by postoperative facial paresis. A sensitivity of 86% and a specificity of 89% were calculated, indicating that A-train occurrence was a highly accurate predictor of poor postoperative facial outcome. A cutoff of 10 s of A-train activity duration has been proposed as a predictor of deterioration of postsurgical N VII function [6, 12]; Liu et al. [4] investigated mean train time as a predictive factor in patients affected by large VSs (>30 mm of diameter) and found that patients with absence of A-train activity were more likely to present good facial function (HB grades 1–2) both in the short term (3–7 days and 3 months after surgery) and in the long term (2 years after surgery) and vice versa. However, further correlation analysis revealed that the predictive effect of mean train time was significant only in the short term; although the authors observed an empiric correlation between longer train time and long-term facial injury, train time could not eventually be proven as a reliable long-term predictive factor.

9.3   Facial Motor Evoked Potentials

The technique of facial motor evoked potentials (FMEPs) has recently been introduced to monitor N VII function and may be considered the most promising method in IOFNM because it surpasses most of the disadvantages of standard techniques [6]. FMEP interpretation is independent from the surgeon's ability to locate the REZ at the brainstem and can facilitate recognition of waveform recordings obtained by free-running EMG [6].

FMEPs consist of stimulation of the area of motor cortex representing facial region and subsequent recording of the responses by the same electrodes and muscle groups used for DES and free-running EMG. Transcranial electrocortical stimulation (TES) is performed intermittently with brainstem auditory evoked potentials and somatosensory evoked potentials (SEPs); it is accomplished using corkscrew-like electrodes inserted in the scalp and positioned at CZ (reference electrode) and C3 or C4 (international 10–20 electroencephalography system) for left- or right-side stimulation, respectively. The responses are detected by bipolar needle electrodes positioned subdermally within the orbicularis oculi and the orbicularis oris muscles. Stimulation is applied contralaterally to the affected side using rectangular pulses, whose number and intensity are not standardized and vary according to different protocols [6].

The presence of muscle MEPs indicates the preservation of all the steps along the motor pathway, i.e., the motor cortex, the corticospinal tract, the alpha motor neurons, the N VII, and the neuromuscular junction. MEP amplitude reduction may be interpreted as a pathological sign; however, several confounding factors are the cause of false-positive responses and may be correlated to the dysfunctions of the motor pathway or to technical issues. Corticospinal tract injury, root or peripheral nerve trauma, stretching, ischemia, or pressure may all be responsible for pathologic MEP amplitude reductions. Neuromuscular blocking agents, stimuli failure, and scalp edema technically interfere with impulse conduction [6]. On the other hand, provided that FMEPs are generated by subpopulations of N VII axons, false-negative results are reported that may be due to minor injuries affecting non-stimulated fibers [6].

A final-to-baseline FMEP amplitude ratio reduction of 50% at the end of the surgery has been identified as a good predictor for postoperative N VII outcome after CPA and skull base surgeries [6, 13]. This criterion was arbitrarily defined by considering the wide variability of FMEP amplitude between patients. Liu et al. [4] studied the role of FMEPs in large (>30 mm) VSs and confirmed that final-to-baseline FMEP amplitude ratios were significantly higher in patients presenting with good N VII function outcome than in those with poor (HB grades 3–6) result at third to seventh day, third month, and second year after surgery.

Acioly et al. [14, 15] hypothesized that intraoperative variation of FMEP amplitude could also correlate with postoperative facial function, even if final-to-baseline amplitude ratios remained above the 50% threshold level. Hence, they investigated event-to-baseline FMEP amplitude ratio and changes in FMEP waveform morphology as predictive factors of immediate and late postoperative facial function [14, 15]. Analysis of correlation coefficients revealed a statistically significant negative correlation of orbicularis oris FMEP amplitude ratio and waveform complexity with the immediate and late postoperative N VII outcome; therefore, greater FMEP amplitude and complexity predicted better N VII function during all surgical stages directly related to tumor manipulation [14, 15]. The study justifies modifications of the surgical strategy based on FMEP amplitude ratio, and wave morphology deterioration as FMEP disappearance has been reported to invariably lead to severe facial palsy and recovery has never been observed [6, 14, 15].

References

1. Vivas EX, Carlson ML, Neff BA, Shepard NT, McCracken DJ, Sweeney AD, et al. Congress of neurological surgeons systematic review and evidence-based guidelines on intraoperative cranial nerve monitoring in vestibular schwannoma surgery. Neurosurgery. 2018;82(2):E44–E6.
2. Amano M, Kohno M, Nagata O, Taniguchi M, Sora S, Sato H. Intraoperative continuous monitoring of evoked facial nerve electromyograms in acoustic neuroma surgery. Acta Neurochir. 2011;153(5):1059–67; discussion 67.
3. Duarte-Costa S, Vaz R, Pinto D, Silveira F, Cerejo A. Predictive value of intraoperative neurophysiologic monitoring in assessing long-term facial function in grade IV vestibular schwannoma removal. Acta Neurochir. 2015;157(11):1991–7; discussion 8.
4. Liu SW, Jiang W, Zhang HQ, Li XP, Wan XY, Emmanuel B, et al. Intraoperative neuromonitoring for removal of large vestibular schwannoma: facial nerve outcome and predictive factors. Clin Neurol Neurosurg. 2015;133:83–9.
5. Mastronardi L, Cacciotti G, Roperto R. Intracanalicular vestibular schwannomas presenting with facial nerve paralysis. Acta Neurochir. 2018;160(4):689–93.
6. Acioly MA, Liebsch M, de Aguiar PH, Tatagiba M. Facial nerve monitoring during cerebellopontine angle and skull base tumor surgery: a systematic review from description to current success on function prediction. World Neurosurg. 2013;80(6):e271–300.
7. Nejo T, Kohno M, Nagata O, Sora S, Sato H. Dorsal displacement of the facial nerve in acoustic neuroma surgery: clinical features and surgical outcomes of 21 consecutive dorsal pattern cases. Neurosurg Rev. 2016;39(2):277–88; discussion 88.
8. Sameshima T, Morita A, Tanikawa R, Fukushima T, Friedman AH, Zenga F, et al. Evaluation of variation in the course of the facial nerve, nerve adhesion to tumors, and postoperative facial palsy in acoustic neuroma. J Neurol Surg B Skull Base. 2013;74(1):39–43.

9. Mastronardi L, Cacciotti G, Roperto R, Di Scipio E, Tonelli MP, Carpineta E. Position and course of facial nerve and postoperative facial nerve results in vestibular schwannoma microsurgery. World Neurosurg. 2016;94:174–80.

10. Bozorg Grayeli A, Kalamarides M, Fraysse B, Deguine O, Favre G, Martin C, et al. Comparison between intraoperative observations and electromyographic monitoring data for facial nerve outcome after vestibular schwannoma surgery. Acta Otolaryngol. 2005;125(10):1069–74.

11. Romstöck J, Strauss C, Fahlbusch R. Continuous electromyography monitoring of motor cranial nerves during cerebellopontine angle surgery. J Neurosurg. 2000;93(4):586–93.

12. Prell J, Rampp S, Romstöck J, Fahlbusch R, Strauss C. Train time as a quantitative electromyographic parameter for facial nerve function in patients undergoing surgery for vestibular schwannoma. J Neurosurg. 2007;106(5):826–32.

13. Matthies C, Raslan F, Schweitzer T, Hagen R, Roosen K, Reiners K. Facial motor evoked potentials in cerebellopontine angle surgery: technique, pitfalls and predictive value. Clin Neurol Neurosurg. 2011;113(10):872–9.

14. Acioly MA, de Aguiar PH, Tatagiba M. Continuous monitoring of evoked facial nerve electromyograms: a new device for an old concept. Acta Neurochir. 2011;153(11):2271–2; author reply 3–4.

15. Acioly MA, Gharabaghi A, Liebsch M, Carvalho CH, Aguiar PH, Tatagiba M. Quantitative parameters of facial motor evoked potential during vestibular schwannoma surgery predict postoperative facial nerve function. Acta Neurochir. 2011;153(6):1169–79.

Hearing Preservation

10

Luciano Mastronardi, Alberto Campione, Ali Zomorodi,
Ettore Di Scipio, Antonio Adornetti,
and Takanori Fukushima

Surgical treatment of vestibular schwannoma (VS) has evolved from a significantly morbid procedure with high mortality to one aimed at tumor resection as well as hearing and facial function preservation. VSs often present with hearing loss and tinnitus. The wider availability of magnetic resonance imaging (MRI) has allowed for early detection of smaller VSs (<2 cm), and therefore more patients present with nearly normal to serviceable hearing. The best therapeutic outcome for small tumors is still under much debate and study. The options for treatment include watchful waiting (WW), stereotactic radiosurgery (SRS), and microsurgical resection (MS) [1].

The guidelines on hearing preservation (HP) outcomes in patients with sporadic VSs by the Congress of Neurological Surgeons (CNS) [2] report specific recommendations about patients counseling as regards the therapeutic options; however, no recommendations are reported about which one should be preferred, and the individual choice by the patients themselves is rather emphasized. The systematic reviews at the basis of the guidelines allowed for the calculation of the probability of maintaining serviceable hearing with WW, SRS, or MS at 2, 5, and 10 years

L. Mastronardi (✉) · A. Campione · A. Adornetti
Department of Neurosurgery, San Filippo Neri Hospital—ASLRoma1, Rome, Italy
e-mail: mastro@tin.it; adornetti@hospitaldevice.it

A. Zomorodi · T. Fukushima
Division of Neurosurgery, Duke University Medical Center, Carolina Neuroscience Institute, Raleigh, NC, USA
e-mail: ali.zomorodi@duke.edu; Fukushima@carolinaneuroscience.com

E. Di Scipio
Department of Neurology and Neurophysiology, San Filippo Neri Hospital—ASLRoma1, Rome, Italy

© Springer Nature Switzerland AG 2019
L. Mastronardi et al. (eds.), *Advances in Vestibular Schwannoma Microneurosurgery*, https://doi.org/10.1007/978-3-030-03167-1_10

following treatment. Tables 10.1 and 10.2 show the data extracted from the guidelines in respect of overall HP probability and HP probability among patients with AAO-HNS class A at baseline, respectively. AAO-HNS (American Academy of Otolaryngology-Head and Neck Surgery) hearing classification [3] is based on the evaluation of both pure tone average (PTA) and speech discrimination score (SDS); class A corresponds to PTA ≤30 dB and SDS ≥70% and class B to PTA ≤50 dB and SDS ≥50%. The same cutoff values are used to define I (good to excellent hearing) and II (serviceable hearing) grade of the Gardner-Robertson Scale [4], respectively; therefore, when "serviceable" hearing is mentioned, AAO-HNS classes A and B are intended.

What emerges from this data is that the HP outcome with MS is most influenced by the selection of eligible patients, as stated in the guidelines on surgical resection for the treatment of VSs by the CNS: HP surgery via the middle fossa or the retrosigmoid approach may be attempted in patients with small tumor size (<1.5 cm) and good preoperative hearing [5].

Golfinos et al. [6] compared outcomes of MS versus SRS in 399 small- and medium-sized VSs (≤2.8 cm of maximum diameter): SRS was associated with better HP and reduced morbidity, whereas facial function was good in both. In conclusion, SRS ensured good tumor control rates, functional hearing, and facial results especially in small VSs, but did not have a curative effect. As a consequence, in case

Table 10.1 Overall probability of maintaining serviceable hearing (AAO-HNS class B at least, according to the definition of "serviceable" by the Gardner-Robertson Scale), data from guidelines [2]

	Watchful waiting	Stereotactic radiosurgery	Microsurgical resection
Early after surgery	–	–	Moderately low probability (>25–50%)
At 2 years	High probability (>75–100%)	Moderately high probability (>50–75%)	Moderately low probability (>25–50%)
At 5 years	Moderately high probability (>50–75%)	Moderately high probability (>50–75%)	Moderately low probability (>25–50%)
At 10 years	Moderately low probability (>25–50%)	Moderately low probability (>25–50%)	Moderately low probability (>25–50%)

Table 10.2 Probability of maintaining serviceable hearing (AAO-HNS class B at least, according to the definition of "serviceable" by the Gardner-Robertson Scale) among patients with AAO-HNS class A at baseline [2]

	Watchful waiting	Stereotactic radiosurgery	Microsurgical resection
Early after surgery	–	–	Moderately high probability (>50–75%)
At 2 years	High probability (>75–100%)	High probability (>75–100%)	Moderately high probability (>50–75%)
At 5 years	Moderately high probability (>50–75%)	Moderately high probability (>50–75%)	Moderately high probability (>50–75%)
At 10 years	Insufficient data available	Moderately low probability (>25–50%)	Moderately low probability (>25–50%)

of failure of such a conservative treatment, a secondary surgical procedure would result in less favorable functional outcomes because of previous radiation therapy.

As regards the sole HP, the apparent advantages of less invasive options like WW and SRS may be in reality problematic in the long term. In fact, observation involves a loss of hearing even in nongrowing tumors [7, 8]; on the other side, radiotherapy leads to no change in hearing in the short term, but progressive and severe loss in the mid-long term has been reported and confirmed by systematic reviews [1, 2, 7, 9, 10]. The amount of these sequelae is not inferior to those produced by primary HPS [11], and for this reason HP should be attempted in every case of serviceable preoperative hearing [1].

The literature about HP surgery is characterized by diffuse heterogeneity in eligibility criteria to select ideal patients, operative techniques, and mean follow-up period. Mazzoni et al. [11] reported a composite series of 322 cases from 1976 to 2009 wherein HP surgery was attempted based on diverse eligibility criteria that changed over the years. The last established criteria proposed by the authors were preoperative AAO-HNS class A, tumor size ≤ 10 mm, and preserved ABR (auditory brainstem response); when the same criteria were retrospectively applied to the entire series, 42 compatible cases were found. Among these, 48% retained AAO-HNS class A hearing, and as many as 83% retained serviceable hearing postoperatively. The authors further stratified the aforementioned 42-patient cohort according to PTA and SDS; they observed that the patients with preoperative PTA ≤ 20 dB and SDS $\geq 80\%$ had 76% of probability of maintaining AAO-HNS class A hearing after HP surgery. It is clear then that the most important predictive factors for a good short-term hearing outcome were the size of the tumor and the preoperative hearing status. Based on radiological observations, the dilation of the internal auditory canal (IAC) caused by the tumor was found to be of prognostic value, too. When the diameter of the canal, as seen in bone window CT, was much larger than on the unaffected side, the cochlear nerve was thinned out by the tumor and more likely to lose function due to dissection.

Yang et al. [12] studied the functional outcomes of retrosigmoid approach; as regards HP, it was retrospectively evaluated within a cohort of patients with tumors of <20 mm of size and AAO-HNS classes A–D of hearing. As far as patients with preoperative serviceable hearing were concerned (a larger and more diverse group compared with that in Mazzoni et al. [11]), 36% of serviceable HP was reported, which raised to 48% after the stratification for tumor size: VSs ≤ 10 mm had a significantly better outcome than larger tumors ($p < 0.05$). In addition, Yang et al. [12] addressed preoperative high-frequency pure tone thresholds as predictive factors better correlated to postoperative PTA than preoperative low frequencies and ABR; the latter, in particular, did not show any correlation with the postoperative hearing function. In 2006, Samii et al. [13] reported total removal by retrosigmoid approach in 98% of 200 cases and HP in 51%. They concluded that total microsurgical removal of small VS (<20 mm) is feasible and curative in one stage, with good preservation of neurological functions, including hearing in patients with preoperative serviceable hearing.

The hearing prognosis of large VSs is controversial and depends on the cutoff used to define them. In their 592-patient series, Wanibuchi et al. [14] reported HP in 53.7% of VSs >2 cm. Di Maio et al. [15] examined 28 cases of VSs ≥3 cm; the authors reported a HP rate of 21.4% which raised to 30.8% if only preoperatively AAO-HNS class A patients were considered. Although the size of the tumors clearly negatively affected the eventual outcome, it was evident how strong the correlation was between preoperative and postoperative hearing status, even in this unfavorable cohort. In addition, Di Maio et al. [15] described two independent factors predictive of better postoperative HP: the presence of a CSF cleft in the IAC as seen on the MRI before surgery and tumor extension anterior to the longitudinal axis of the IAC <35% of total tumor volume.

The short-term HP is often interpreted as a measure of the surgeon's ability to anatomically preserve the cochlear nerve; however, immediate postoperative anacusis or significant hearing deterioration have been reported also in cases of apparently intact cochlear nerves. In such circumstances, the contribution by the cochlea itself to elicitation and transmission of electrical impulses must be taken into consideration. Indeed, a loss of hair cell function may occur, and the most likely explanation for this is that the internal auditory artery supplying both the cochlea and the cochlear nerve is exposed to damage during dissection, as complete preservation of this artery is extremely difficult when the eighth nerve is severed [16]. Even more complex is the widely reported [1, 11, 17, 18] case of long-term hearing function decay after successful postoperative HP. Strauss et al. [19] suggest that surgical manipulations can initiate disturbances in microcirculation in the endoneurial vasa nervorum. Disturbed microcirculation can lead to massive releases of glutamate during and after nerve ischemia, which in turn can cause an influx of calcium into the damaged neurons, resulting in cell death. Mechanical or microvascular trauma at the site of tumor removal can be expected to initially affect only the distal portion of the cochlear nerve (which reflects in the delayed loss of components of ABR beyond wave I), followed by a gradual loss of wave I as neural degeneration progresses laterally.

The most recent meta-analysis about long-term hearing outcome after MS as primary treatment by Ahsan et al. [1] revealed that if serviceable hearing is retained at early postoperative visit, the chance of preserving hearing at >5 years is excellent. The overall HP rate immediately after surgery was esteemed as ranging between 50 and 70%. The mean hearing retention rate in the long term being 70%, 35–49% of all patients undergoing HP surgery would continue to maintain serviceable hearing (AAO-HNS class A or B) at 5 years after surgery. This figure is in line with the long-term hearing stability for VSs managed conservatively, which ranges between 41 and 57% [1]. Meanwhile, after SRS, HP has been reported to be about 74 and 44.5% at 3 and 10 years, respectively [8]. Ahsan et al. [1] also reported that those who maintained SDS ≥89% at the early postoperative follow-up had better long-term HP, which justifies the choice of MS as a primary treatment even in cases of VSs presenting with normal hearing function. The meta-analysis confirmed that only preoperative and postoperative PTA are associated with long-term hearing preservation, a correlation that had previously been found by Nakamizo et al. [18].

The latter authors analyzed the long-term hearing outcomes at a mean follow-up of 5 years in patients with unilateral VS treated via the retrosigmoid approach; during the postoperative follow-up, PTA was reevaluated within 6 months after surgery in seven patients. In the two patients whose PTA deteriorated ≥5 dB, their PTA worsened ≥15 dB at the final follow-up compared to the immediate postoperative PTA. In the remaining five patients whose PTA deteriorated <5 dB in 6 months after surgery, PTA was maintained within a 15-dB deterioration at the final follow-up ($p = 0.04$). As a result, the deterioration of PTA in the early postoperative period may help to predict the long-term outcomes of hearing function.

Even longer follow-up of at least 6 years (range 6–21 years) was conducted by Mazzoni et al. [17]. In their study, 87% of the patients who had serviceable hearing early after MS maintained the same hearing function at the final follow-up; this means that a 13% hearing decay was reported, perfectly in line with the previous study on the same series by these authors [11]. Seen in a global perspective, of 189 patients with AAO-HNS class A or B before surgery, 54 retained the same hearing quality in the short term and 47 in the long term, thus yielding rates as high as 29% and 25%, respectively. Although positive, the authors considered that such results are not better than those obtained in the long-term SRS; however, it should be noted that the patients monitored for this study were operated on using different and looser eligibility criteria for HP surgery than those proposed by the same group later [11]. Thus, even better long-term results may be expected.

10.1   Intraoperative Cochlear Nerve Monitoring

According to the latest guidelines by the CNS [20], intraoperative cochlear nerve monitoring (IOCNM) should be used during VS surgery when HP is attempted, in case of measurable preoperative hearing levels and tumors <1.5 cm; in respect of the preferable technique of IOCNM, there is insufficient evidence to conclude whether direct monitoring of the eighth cranial nerve is superior to the use of far-field ABRs [20].

ABR is a far-field evoked potential that requires a dedicated apparatus. After induction of general anesthesia, a soft ear mold attached with a 12-inch plastic tube is placed and sealed inside the ear canal. Surface electrodes are placed at the vertex (Cz) and on each earlobe (A1 and A2). Two channels are used, A1–Cz and A2–Cz, to elicit and collect responses also on the unaffected side [21, 22]. A brief click or tone is delivered to the affected side at a sound pressure of 90–100 dB and at a rate of 31–51 Hz. Contralateral ear is usually masked by white noise at 50 dB [21, 22]. Baseline responses for each ear are recorded before the beginning of surgery and are used as baselines throughout the case. The classic ABR comprises 5–7 peaks, all occurring within 10 ms of the click; the first 5 peaks (waves I–V) are the most important in clinical practice. Waves IV and V are generated at the upper pons and lower midbrain. Wave V tends to be the most robust and is the most closely monitored during surgery. The surgeon is alerted when the change in latency of wave V exceeds 0.5 ms or if there is a change in or disappearance of any wave [23].

The main limit of ABR is the amplitude of the responses, less than 1 μV, that needs summation and a long data acquisition time to achieve an adequate signal-to-noise ratio. As a result, the technique of ABR has poor temporal resolution and is susceptible to disruption by various intraoperative factors, including dural opening, saline irrigation of surgical field, surgical microscope, high-speed drill, and ultrasonic aspirator [27]. Despite these disadvantages, in 2016, Hummel et al. [24] observed that ABR is a predictor of postoperative cochlear nerve function. ABR quality after 60% tumor removal was independently significant for hearing outcome, possibly resulting from progressive damage to cochlear nerve during resection or because dissection of the tumor capsule from the nerves took place in the final phase of the procedures [24].

CE-Chirp® ABR represents a recent development of classical ABR (Fig. 10.1). CE-Chirp® is a new acoustic stimulus used in newborn hearing testing, designed to provide enhanced neural synchronicity and faster detection of larger amplitude wave V. CE-Chirp® acoustic stimulus, developed by Claus Elberling, has the same spectrum and the same calibration as a usual square wave click stimulus. Acoustic energy from the CE-Chirp® stimulus reaches all regions of the cochlea at approximately the same time [25, 26]. The difference lies in the presentation timing of the low-, mid-, and high-frequency components of acoustic stimuli. This change in the stimulus presentation offsets the mechanics of the cochlea's traveling wave and results in an ABR waveform of increased amplitude than the corresponding click ABR in normal hearing subjects [25, 26] (Fig. 10.2).

In their preliminary study, Mastronardi et al. [22] observed that classical ABR needed a series of about 1000 stimuli, in all patients, to evoke a clear and

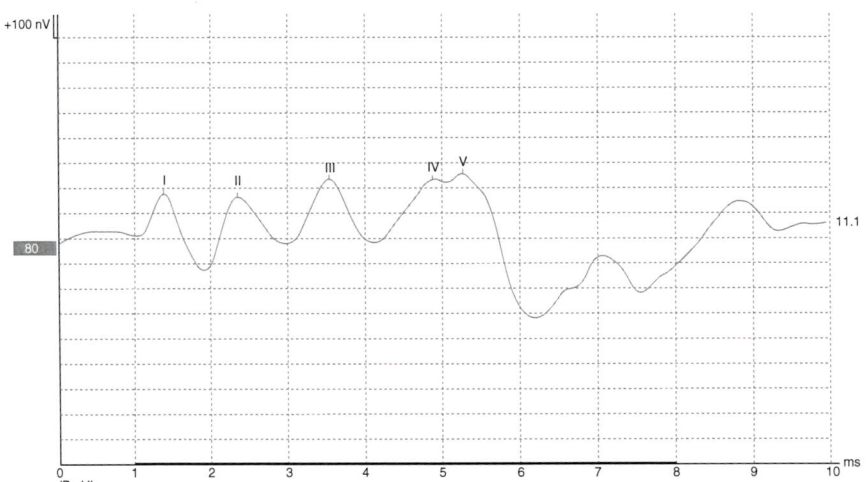

Fig. 10.1 Classical ABR. For journal content: *Reprinted by permission from Springer Customer Service Center GmbH: Springer Nature, Neurosurgical Review, CE-Chirp® ABR in cerebellopontine angle surgery neuromonitoring: technical assessment in four cases, Ettore Di Scipio, Luciano Mastronardi, 2015*

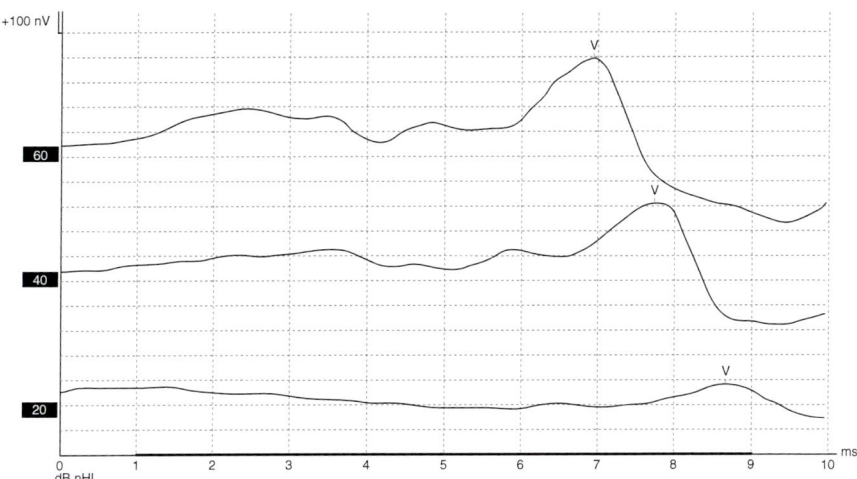

Fig. 10.2 CE-Chirp® ABR at different sound pressures. Compared to Fig. 10.1, wave V shown in this graph reaches a higher intensity (measured in nV) even at lower sound pressures. *Reprinted by permission from Springer Customer Service Centre GmbH: Springer Nature, Neurosurgical Review, CE-Chirp® ABR in cerebellopontine angle surgery neuromonitoring: technical assessment in four cases, Ettore Di Scipio, Luciano Mastronardi, 2015*

monitorable V-wave. Using CE-Chirp® ABR, a series of about 600 stimuli was sufficient, thus reducing the time needed for a successful stimulation; in addition, time analysis was 10 s per sweep, which enabled the monitoring team to alert the surgeon to any change or disappearance of wave V online. The same authors conducted a study to evaluate postoperative HP with reference to tumor size in patients operated on with level-specific (LS)-CE-Chirp® ABR monitoring. Twenty-five patients with preoperative AAO-HNS class A–B hearing were selected and divided into two groups based on tumor size: group A (≤2 cm) and group B (>2 cm). HP rate was 52%, with significant differences in relation to size: 61.5% group A and 41.7% group B ($p = 0.014$).

Yamakami et al. [27] removed small VSs with retrosigmoid approach using a newly designed intracranial electrode enabling continuous monitoring of cochlear nerve compound action potentials (CNAPs). CNAP is a near-field evoked potential obtained from an intracranial electrode placed directly on the cisternal cochlear nerve. In the electrode, a small tuft of cotton is secured on the tip of a fine, malleable urethane-coated wire. According to the technique reported in their study [27], after dural incision the cisternal cochlear nerve was identified as soon as possible. The tip of the intracranial electrode was placed on the cochlear nerve near the root entry zone proximal to the CPA tumor and thereafter was covered by a small surgical cotton pledget to hold and segregate the electrode from the operative field. After the electrode placement and prior to starting microsurgical tumor removal, the baseline CNAP was recorded. CNAPs were monitored continuously throughout intradural microsurgery.

CNAP is the summation of individual nerve fiber signals in the cochlear nerve. Tumor compression of the cochlear nerve causes the conduction block of individual nerve fibers and the desynchronization of electrical signals. The desynchronization itself results in a decreased amplitude or even in the disappearance of CNAPs. Surgical decompression of the cochlear nerve can dissolve the desynchronization, allowing for increased response amplitude after tumor removal [27]. Thus, microsurgical maneuvers induce dynamic changes in CNAPs' morphology and intensity.

In 44 VSs with maximal diameter $\leq$1.5 cm, Yamakami et al. [27–29] observed postoperative serviceable hearing in 72%, concluding that reliable monitoring was more frequently provided by CNAP than by ABR evoked by classical square-wave click stimuli (66% vs. 32%, $p < 0.01$) [27–29] and had better rates of HP [29]. However, HP rates of 61.5% obtained by Mastronardi et al. [22] in VSs $\leq$2 cm did not seem very different: this might suggest that, while CNAP has a number of advantages over classical ABR, CE-Chirp® ABR may allow for comparable results without the risk of frequent displacement associated to the electrode employed in CNAP. Further studies directly comparing the two techniques are needed.

References

1. Ahsan SF, Huq F, Seidman M, Taylor A. Long-term hearing preservation after resection of vestibular schwannoma: a systematic review and meta-analysis. Otol Neurotol. 2017;38(10):1505–11.
2. Carlson ML, Vivas EX, McCracken DJ, Sweeney AD, Neff BA, Shepard NT, et al. Congress of Neurological Surgeons Systematic Review and Evidence-Based Guidelines on Hearing Preservation Outcomes in Patients With Sporadic Vestibular Schwannomas. Neurosurgery. 2018;82(2):E35–E9.
3. Committee on Hearing and Equilibrium guidelines for the evaluation of hearing preservation in acoustic neuroma (vestibular schwannoma). American Academy of Otolaryngology-Head and Neck Surgery Foundation, INC. Otolaryngol Head Neck Surg. 1995;113(3):179–80.
4. Gardner G, Robertson JH. Hearing preservation in unilateral acoustic neuroma surgery. Ann Otol Rhinol Laryngol. 1988;97(1):55–66.
5. Hadjipanayis CG, Carlson ML, Link MJ, Rayan TA, Parish J, Atkins T, et al. Congress of neurological surgeons systematic review and evidence-based guidelines on surgical resection for the treatment of patients with vestibular schwannomas. Neurosurgery. 2018;82(2):E40–E3.
6. Golfinos JG, Hill TC, Rokosh R, Choudhry O, Shinseki M, Mansouri A, et al. A matched cohort comparison of clinical outcomes following microsurgical resection or stereotactic radiosurgery for patients with small- and medium-sized vestibular schwannomas. J Neurosurg. 2016;125(6):1472–82.
7. Hoa M, Drazin D, Hanna G, Schwartz MS, Lekovic GP. The approach to the patient with incidentally diagnosed vestibular schwannoma. Neurosurg Focus. 2012;33(3):E2.
8. Stangerup SE, Thomsen J, Tos M, Cayé-Thomasen P. Long-term hearing preservation in vestibular schwannoma. Otol Neurotol. 2010;31(2):271–5.
9. Patnaik U, Prasad SC, Tutar H, Giannuzzi AL, Russo A, Sanna M. The long-term outcomes of wait-and-scan and the role of radiotherapy in the management of vestibular schwannomas. Otol Neurotol. 2015;36(4):638–46.
10. Prasad SC, Patnaik U, Grinblat G, Giannuzzi A, Piccirillo E, Taibah A, et al. Decision making in the wait-and-scan approach for vestibular schwannomas: is there a price to pay in terms of hearing, facial nerve, and overall outcomes? Neurosurgery. 2018;83(5):858–70.

11. Mazzoni A, Biroli F, Foresti C, Signorelli A, Sortino C, Zanoletti E. Hearing preservation surgery in acoustic neuroma. Slow progress and new strategies. Acta Otorhinolaryngol Ital. 2011;31(2):76–84.
12. Yang J, Grayeli AB, Barylyak R, Elgarem H. Functional outcome of retrosigmoid approach in vestibular schwannoma surgery. Acta Otolaryngol. 2008;128(8):881–6.
13. Samii M, Gerganov V, Samii A. Improved preservation of hearing and facial nerve function in vestibular schwannoma surgery via the retrosigmoid approach in a series of 200 patients. J Neurosurg. 2006;105(4):527–35.
14. Wanibuchi M, Fukushima T, Friedman AH, Watanabe K, Akiyama Y, Mikami T, et al. Hearing preservation surgery for vestibular schwannomas via the retrosigmoid transmeatal approach: surgical tips. Neurosurg Rev. 2014;37(3):431–44; discussion 44.
15. Di Maio S, Malebranche AD, Westerberg B, Akagami R. Hearing preservation after microsurgical resection of large vestibular schwannomas. Neurosurgery. 2011;68(3):632–40; discussion 40.
16. Babbage MJ, Feldman MB, O'Beirne GA, Macfarlane MR, Bird PA. Patterns of hearing loss following retrosigmoid excision of unilateral vestibular schwannoma. J Neurol Surg B Skull Base. 2013;74(3):166–75.
17. Mazzoni A, Zanoletti E, Calabrese V. Hearing preservation surgery in acoustic neuroma: long-term results. Acta Otorhinolaryngol Ital. 2012;32(2):98–102.
18. Nakamizo A, Mori M, Inoue D, Amano T, Mizoguchi M, Yoshimoto K, et al. Long-term hearing outcome after retrosigmoid removal of vestibular schwannoma. Neurol Med Chir (Tokyo). 2013;53(10):688–94.
19. Strauss C, Bischoff B, Neu M, Berg M, Fahlbusch R, Romstöck J. Vasoactive treatment for hearing preservation in acoustic neuroma surgery. J Neurosurg. 2001;95(5):771–7.
20. Vivas EX, Carlson ML, Neff BA, Shepard NT, McCracken DJ, Sweeney AD, et al. Congress of neurological surgeons systematic review and evidence-based guidelines on intraoperative cranial nerve monitoring in vestibular schwannoma surgery. Neurosurgery. 2018;82(2):E44–E6.
21. Di Scipio E, Mastronardi L. CE-Chirp® ABR in cerebellopontine angle surgery neuromonitoring: technical assessment in four cases. Neurosurg Rev. 2015;38(2):381–4; discussion 4.
22. Mastronardi L, Di Scipio E, Cacciotti G, Roperto R. Vestibular schwannoma and hearing preservation: usefulness of level specific CE-Chirp ABR monitoring. A retrospective study on 25 cases with preoperative socially useful hearing. Clin Neurol Neurosurg. 2018;165:108–15.
23. Youssef AS, Downes AE. Intraoperative neurophysiological monitoring in vestibular schwannoma surgery: advances and clinical implications. Neurosurg Focus. 2009;27(4):E9.
24. Hummel M, Perez J, Hagen R, Gelbrich G, Ernestus RI, Matthies C. Auditory monitoring in vestibular schwannoma surgery: intraoperative development and outcome. World Neurosurg. 2016;96:444–53.
25. Elberling C, Don M. Auditory brainstem responses to a chirp stimulus designed from derived-band latencies in normal-hearing subjects. J Acoust Soc Am. 2008;124(5):3022–37.
26. Elberling C, Don M, Cebulla M, Stürzebecher E. Auditory steady-state responses to chirp stimuli based on cochlear traveling wave delay. J Acoust Soc Am. 2007;122(5):2772–85.
27. Yamakami I, Yoshinori H, Saeki N, Wada M, Oka N. Hearing preservation and intraoperative auditory brainstem response and cochlear nerve compound action potential monitoring in the removal of small acoustic neurinoma via the retrosigmoid approach. J Neurol Neurosurg Psychiatry. 2009;80(2):218–27.
28. Yamakami I, Oka N, Yamaura A. Intraoperative monitoring of cochlear nerve compound action potential in cerebellopontine angle tumour removal. J Clin Neurosci. 2003;10(5):567–70.
29. Yamakami I, Ushikubo O, Uchino Y, Kobayashi E, Saeki N, Yamaura A, et al. [Intraoperative monitoring of hearing function in the removal of cerebellopontine angle tumor: auditory brainstem response and cochlear nerve compound action potential]. No Shinkei Geka. 2002;30(3):275–82.

Usefulness of Laser and Ultrasound Aspirator

11

Luciano Mastronardi, Alberto Campione, Ali Zomorodi,
Raffaelino Roperto, Guglielmo Cacciotti,
and Takanori Fukushima

11.1 Laser

Lasers proved to be well-established instruments in different surgical fields for more than 40 years [1–3]. The rationale for laser use in tumor resection is both to allow for "no-touch" cutting and for tissue debulking, with hemostatic benefit [4]. Laser surgery, in general, has shown various advantages, such as reduction of mechanical trauma and intraoperative bleeding [5–7]. Three types of laser have been successfully used in vestibular schwannoma (VS) surgery: potassium titanyl phosphate (KTP-532), CO_2, and the novel 2μ-thulium lasers.

KTP-532 pulsed-wave laser has a wavelength of 532 nm, which renders it absorbable by hemoglobin but not by water; it was most used until the development of flexible CO_2 continuous-wave laser fiber in 2005 [8–10]. Nissen et al. [11] presented a series of 111 patients in whom KTP-532 laser surgery was used in VSs; the authors reported that laser dissection did not result in deleterious neurological sequelae or laser-specific complications. In addition, the facial functional outcome was reported as in line with the literature describing non-laser dissection techniques; according to the House-Brackmann (HB) grading system [12], 90.2% of small tumors, 72.2% of medium tumors, and 75% of large tumors achieved satisfactory (HB 1–2) functional results [11].

Stellar et al. reported use of a CO_2 laser for resection of an intracranial tumor in 1970 [13]. The CO_2 laser has particular advantages in surgery. Its infrared wavelength (10.6 μm) penetrates water very poorly, confining its area of action to the

L. Mastronardi (✉) · A. Campione · R. Roperto · G. Cacciotti
Department of Neurosurgery, San Filippo Neri Hospital—ASLRoma1, Rome, Italy
e-mail: mastro@tin.it

A. Zomorodi · T. Fukushima
Division of Neurosurgery, Duke University Medical Center, Carolina Neuroscience Institute, Raleigh, NC, USA
e-mail: ali.zomorodi@duke.edu; Fukushima@carolinaneuroscience.com

© Springer Nature Switzerland AG 2019
L. Mastronardi et al. (eds.), *Advances in Vestibular Schwannoma Microneurosurgery*, https://doi.org/10.1007/978-3-030-03167-1_11

surface of the biological structures being dissected and minimizing collateral tissue damage [2, 14, 15]. Moreover, it is a continuous-wave laser energy which avoids the explosive effects of pulsed-wave lasers and allows accurate cutting and vaporization by using focused beams, without the need for tissue handling or retraction [2]. In the pre-2005 era, CO_2 laser energy could only be transmitted through a housing with mirrors and bulky articulating arms which constituted a considerable ergonomic and practical disadvantage. Indeed, all materials available at the time for fiber optic transmission were opaque to light in the infrared spectrum at 10.6 μm; thus, fiber optic cables could not be used [2]. Eiras et al. [16] presented their results after the resection of 12 cases of giant VSs operated on with microsurgical technique and CO_2 laser. In spite of successful facial nerve preservation, they reported that the laser technique required a longer duration of surgery than the traditional microsurgical technique (6.1 h vs. 5.5 h), which could be due to the impractical design of the laser device itself [16]. After 2005, the new devices offered the possibility of guiding the CO_2 laser beam via small and variable handpieces for direct microsurgical application. Thus, it could be demonstrated that this laser is more precise and less damaging to the surrounding tissue than conventional bipolar cauterization [15].

Scheich et al. [7] analyzed the results of microsurgery in VSs with assistance of a flexible CO_2 laser fiber (Omniguide®, FELS 30A, Omniguide Inc., Cambridge, MA, USA) using the middle fossa (MF) approach. A group of 20 consecutive patients with VS stage T1/T2 and AAO-HNS class of hearing A–B [17] was operated on with laser technique (laser group, LG), and the results were compare to those obtained in a control group (CG). All the patients of both groups had normal facial function (HB 1) preoperatively. At 1 week after surgery, 70% of patients retained HB 1 facial function in both groups; at 3 months of follow-up, 100% and 95% of the patients had completely recovered (HB 1) in LG and CG, respectively. All the patients of both groups had serviceable hearing preoperatively, according to the definition of "serviceable hearing" by the Gardner-Robertson scale [18]. Hearing preservation rate was 72% in LG and 82% in CG, with no statistically significant difference between the groups. No statistical differences in mean operation time (from skin incision to suture) were reported. The authors concluded that the use of a handheld flexible CO_2 laser fiber in VS microsurgery is safe and leads to functional outcomes comparable with those obtained in conventionally treated patient. Laser was deemed particularly useful for resection especially of "difficult" (e.g., highly vascularized) tumors [7].

Schwartz et al. [4] reviewed 41 cases of medium- and large-sized VSs that were treated using the flexible CO_2 laser fiber (Omniguide®, FELS-25A, ARC Laser GmbH) during retrosigmoid (RS) or translabyrinthine (TL) approaches. Time of resection and blood loss were compared to a control group and showed no statistically significant differences. Preoperatively, 97.6% of patients had normal facial function (HB 1), which was preserved in 70.7% in the first postoperative days and ultimately in 92.7% at the last follow-up visit. Of the four patients undergoing attempted hearing preservation surgery with the use of the CO_2 laser via the RS approach, two (50%) had preserved hearing at the preoperative level (one each in AAO-HNS class A and class B). In summary, cranial nerve preservation compared favorably to that in other reports [19–21]. As regards the role of laser during surgery, the authors found that it functioned best as a cutting tool—with several advantages over microscissors—rather than a vaporizing tool. In addition, tumor could be cut while simultaneously being retracted centrally by a suction device held in the non-dominant hand. This avoided the typical pushing action of microscissors and allowed for the resection of larger pieces of tumor in fewer steps. No-touch tumor resection near the internal acoustic meatus also helped avoid traction on cranial nerves. Most importantly, the authors suggested that the depths of the field always be kept bathed in saline solution, thus protecting submerged structures [4].

2μ-Thulium continuous-wave laser has a wavelength of 2 micron; this allows for excess laser radiation to be absorbed by the irrigation so that it does not affect tissue more than 3 mm away from the tip of the fiber. Tissue damage is limited to 0.2–1.0 mm around, and the minimal width of the fiber allows for perfect visualization and control of the surgical field. 2μ-Thulium laser already showed to be a promising device in the surgery of intracranial meningiomas [22], especially for debulking, shrinking, and coagulating the mass and its basal implant.

Mastronardi et al. analyzed the results of 2μ-thulium-assisted VS microsurgery in two studies [5, 6], the latest of which [5] featured a laser group (LG) of 37 patients and a control group (CG) of 44 patients. In LG, the capsule incision and tumor debulking were performed with handheld 2μ-thulium flexible laser fiber (RevoLix™, Lisa Laser Products, Katlenburg-Lindau, Germany). Standard 0.9% saline solution irrigation was used as a cooling agent. The fiber was used for cutting, vaporizing, and coagulating the capsule and the intracapsular mass, in combination with bipolar forceps, microscissors, and ultrasonic aspirator. Following tumor debulking, the remaining tumor capsule was removed with standard microsurgical tools [5, 6]. Mean operation time changed in relation to size of tumor; instead, no significant differences were reported between LG and CG. As regards facial function, all the patients were considered as HB 1 preoperatively, except five in LG and three in CG. The day after surgery, a normal face was observed in 38.9 and 61.4% of patients in LG and CG, respectively. The lower rate in LG may be due to larger mean size of tumor than in CG, although the difference did not reach statistical significance. At 6 months after surgery, HB 1 facial nerve function was observed with nearly the same rate in both groups (91.7% in LG vs. 93.2% in CG). Preoperatively, 14 patients in LG and 22 patients in CG had serviceable hearing (AAO-HNS class A–B). The hearing preservation rate was 78.6% in LG and 68.2% in CG, with no statistically significant difference. The cranial nerve preservation rates were similar in both groups and comparable to other reports [4, 7, 21, 23]. Therefore, the use of 2μ-thulium handheld flexible laser fiber in VS microsurgery was eventually safe and subjectively facilitated tumor resection especially in "difficult" conditions (e.g., highly vascularized and hard tumors) [5], as already noticed by Scheich et al. [7].

11.2   Ultrasound Aspirator

The ultrasonic aspirator (USA) was first developed for the removal of dental plaque, in 1947. Flamm et al. [24] tested this device on animal brains and first applied it to neurosurgical procedures to remove meningiomas and VSs, in 1978. Design modifications and adjustments to novel surgical techniques continued throughout the 1980s, and the USA became the most common complex adjunct instrument used in the excision of brain tumors [25]. In 1999, a first prototype of what would then become the standard USA was designed by Sawamura et al.; it represented the shift toward more maneuverable devices incorporating electronically controlled irrigation and suction systems without the need for constant cooling [25].

The essential structure of an USA is composed of a handpiece and a probe attached to it; many types of probes exist and differ in the shape of their tips. Two types of handpieces were originally available: the magnetostrictive system, less efficient and prone to overheating because of its coil resistance; and the electrostrictive system, wherein a highly efficient piezoelectric ceramic transducer converted electrical energy into longitudinal mechanical vibrations. As electrostrictive systems did not require any cooling device and were smaller and thus more comfortable to handle, they became and remain the core basis of the modern USA [25]. Vibrations generated by the piezoelectric transducer result in peaks of high and low pressure delivered on the targeted tissue. Cells expand under negative pressure and peaks of high pressure cause them to burst. The process is selective because soft tissues with high water content are more susceptible to cavitation. On the other hand, collagen and elastin fibers tend to vibrate in resonance with the acoustic vibrations. Dedicated cutting ablation tips are required to overcome the phenomenon of resonance in fibrotic or calcified tissues: such tips break the collagen bonds and allow for cavitation.

The USA is used for tissue fragmentation; it was first employed for debulking of gliomas and tumors of the posterior fossa, including VSs. It replaced the microtumor forceps and the cautery (as used in traditional cautery-suction technique) that would cause a significant degree of traction and transmitted movement to the adjacent neurovascular structures [25]. The main disadvantages of older models of Cavitron ultrasonic surgical aspirator (CUSA) were the risk of indirect injury to the cranial nerves during skull base surgery [26] and the easy laceration of nerves and small arteries previously stretched and compressed by tumor growth. However, Sawamura et al. reported no risk of traction injury using their newly designed needle-type probe, which was eventually associated with a smaller risk of perforation of the nerves and the arachnoid membrane, compared with other probes [25]. Thus, modern USAs are supposed to leave the neurovascular structures around the surgical site largely untouched.

As reported by Epstein [27], the CUSA lacked primary hemostatic properties and required conventional hemostatic techniques during tissue fragmentation. In 2000, Kanzaki et al. reported the good hemostatic properties of an ultrasonic activated scalpel: its vibrating blade coupled with tissue proteins mechanically denatured them to form a sticky coagulum that sealed blood vessels (a mechanism known as "coaptive coagulation") [28]. As far as neurovascular preservation is concerned, in their 15-patient series, facial nerve preservation rate was significantly higher than that obtained in patients on whom the ultrasonic activated scalpel was not used ($p < 0.01$) [28].

The use of the USA for meatal bone removal has been recently introduced in routine skull base surgery as an alternative to the neurotologic drill. Previous cadaveric studies [29, 30] described such technique in 2015, and Modest et al. first presented a series of 55 patients who underwent meatal bone removal using the USA during RS craniotomy for VS resection in 2016 [31]. Cadaveric studies by Weber et al. and Golub et al. show that the USA disperses up to 25 times less bone dust than the traditional neurotologic drill [30] and performs bone removal in approximately the same amount of time [29]. Theoretically, the decrease in bone dust dispersion during drilling of internal auditory canal (IAC) could reduce the incidence of postoperative headache. Indeed, several authors have theorized that chemical meningitis due to intracranial bone dust dispersion is a chief reason why the prevalence of postoperative headache is greater following RS craniotomy compared to TL and MF approaches as the RS craniotomy is the only approach that requires intradural drilling of bone to access IAC [32, 33].

Modest et al. reported no injuries to jugular bulb, cranial nerves, vasculature, or cerebellar tissue due to the USA. The time for meatal bone removal using the USA was similar to that of conventional drilling, as expected [29, 31]. However, still considerable bone particle dissemination was reported due to irrigation and cerebrospinal fluid, but—in the authors' opinion—particles produced by the USA were larger than those of a conventional drill and therefore easier to remove from the posterior fossa [31]. As regards facial nerve preservation, temporary (<6 months) facial nerve weakness (HB 2–5) was reported in 11% of patients; 9% were reported to have light facial nerve weakness (HB 2–3) at last follow-up (>6 months) [31]. These results confirmed the safety of the USA as previously reported by Ito et al. [34]. Twelve patients of the series were selected as good candidates for hearing preservation surgery (tumor size <1.5 cm and AAO-HNS class A–B of hearing): 50% of them had successful hearing preservation, and no evidence of acoustic injury due to the USA was reported in the whole series [31], confirming previous reports as well [34, 35]. At last follow-up (>6 months), 15% of patients reported ongoing headache requiring prescription medication [31]. Postoperative headache prevalence after RS craniotomy performed with neurotologic drills ranges from 17 to 80% in the literature [32, 33]. However, the absence of a control group in the study by Modest et al. does not allow to conclude whether the USA eventually exerts a protective effect against postoperative headache.

In conclusion, the USA is ergonomic, safe, and suitable for both bone and tumor removal, obviating the need for a separate drill setup. However, care must be taken in controlling suction, irrigation, and power settings to obtain ideal results and avoid any damage to surrounding dura and soft tissues; the senior authors of this book suggest the configuration Power:50, Suction:5, Irrigation:5 for VS debulking and increasing power settings for IAC unroofing.

References

1. Gardner G, Robertson JH, Clark WC, Bellott AL, Hamm CW. Acoustic tumor management—combined approach surgery with CO2 laser. Am J Otol. 1983;5(2):87–108.
2. Ryan RW, Spetzler RF, Preul MC. Aura of technology and the cutting edge: a history of lasers in neurosurgery. Neurosurg Focus. 2009;27(3):E6.
3. Tew JM, Tobler WD. Present status of lasers in neurosurgery. Adv Tech Stand Neurosurg. 1986;13:3–36.
4. Schwartz MS, Lekovic GP. Use of a flexible hollow-core carbon dioxide laser for microsurgical resection of vestibular schwannomas. Neurosurg Focus. 2018;44(3):E6.
5. Mastronardi L, Cacciotti G, Roperto R, Tonelli MP, Carpineta E, How I. Do it: the role of flexible hand-held 2μ-thulium laser fiber in microsurgical removal of acoustic neuromas. J Neurol Surg B Skull Base. 2017;78(4):301–7.
6. Mastronardi L, Cacciotti G, Scipio ED, Parziale G, Roperto R, Tonelli MP, et al. Safety and usefulness of flexible hand-held laser fibers in microsurgical removal of acoustic neuromas (vestibular schwannomas). Clin Neurol Neurosurg. 2016;145:35–40.
7. Scheich M, Ginzkey C, Harnisch W, Ehrmann D, Shehata-Dieler W, Hagen R. Use of flexible CO_2 laser fiber in microsurgery for vestibular schwannoma via the middle cranial fossa approach. Eur Arch Otorhinolaryngol. 2012;269(5):1417–23.
8. Hart SD, Maskaly GR, Temelkuran B, Prideaux PH, Joannopoulos JD, Fink Y. External reflection from omnidirectional dielectric mirror fibers. Science. 2002;296(5567):510–3.
9. Ibanescu M, Fink Y, Fan S, Thomas EL, Joannopoulos JD. An all-dielectric coaxial waveguide. Science. 2000;289(5478):415–9.
10. Temelkuran B, Hart SD, Benoit G, Joannopoulos JD, Fink Y. Wavelength-scalable hollow optical fibres with large photonic bandgaps for CO2 laser transmission. Nature. 2002;420(6916):650–3.
11. Nissen AJ, Sikand A, Welsh JE, Curto FS. Use of the KTP-532 laser in acoustic neuroma surgery. Laryngoscope. 1997;107(1):118–21.
12. House JW, Brackmann DE. Facial nerve grading system. Otolaryngol Head Neck Surg. 1985;93(2):146–7.
13. Stellar S, Polanyi TG, Bredemeier HC. Experimental studies with the carbon dioxide laser as a neurosurgical instrument. Med Biol Eng. 1970;8(6):549–58.
14. Cerullo LJ, Mkrdichian EH. Acoustic nerve tumor surgery before and since the laser: comparison of results. Lasers Surg Med. 1987;7(3):224–8.
15. Ryan RW, Wolf T, Spetzler RF, Coons SW, Fink Y, Preul MC. Application of a flexible CO(2) laser fiber for neurosurgery: laser-tissue interactions. J Neurosurg. 2010;112(2):434–43.
16. Eiras J, Alberdi J, Gomez J. Laser CO2 in the surgery of acoustic neuroma. Neurochirurgie. 1993;39(1):16–21; discussion 21–3.
17. Committee on Hearing and Equilibrium guidelines for the evaluation of hearing preservation in acoustic neuroma (vestibular schwannoma). American Academy of Otolaryngology-Head and Neck Surgery Foundation, INC. Otolaryngol Head Neck Surg. 1995;113(3):179–80.
18. Gardner G, Robertson JH. Hearing preservation in unilateral acoustic neuroma surgery. Ann Otol Rhinol Laryngol. 1988;97(1):55–66.
19. Ben Ammar M, Piccirillo E, Topsakal V, Taibah A, Sanna M. Surgical results and technical refinements in translabyrinthine excision of vestibular schwannomas: the Gruppo Otologico experience. Neurosurgery. 2012;70(6):1481–91; discussion 91.
20. Nonaka Y, Fukushima T, Watanabe K, Friedman AH, Sampson JH, Mcelveen JT, et al. Contemporary surgical management of vestibular schwannomas: analysis of complications and lessons learned over the past decade. Neurosurgery. 2013;72(2 Suppl Operative):ons103–15; discussion ons15.
21. Samii M, Gerganov V, Samii A. Improved preservation of hearing and facial nerve function in vestibular schwannoma surgery via the retrosigmoid approach in a series of 200 patients. J Neurosurg. 2006;105(4):527–35.

22. Passacantilli E, Antonelli M, D'Amico A, Delfinis CP, Anichini G, Lenzi J, et al. Neurosurgical applications of the 2-μm thulium laser: histological evaluation of meningiomas in comparison to bipolar forceps and an ultrasonic aspirator. Photomed Laser Surg. 2012;30(5):286–92.
23. Wanibuchi M, Fukushima T, Friedman AH, Watanabe K, Akiyama Y, Mikami T, et al. Hearing preservation surgery for vestibular schwannomas via the retrosigmoid transmeatal approach: surgical tips. Neurosurg Rev. 2014;37(3):431–44; discussion 44.
24. Flamm ES, Ransohoff J, Wuchinich D, Broadwin A. Preliminary experience with ultrasonic aspiration in neurosurgery. Neurosurgery. 1978;2:240–5.
25. Sawamura Y, Fukushima T, Terasaka S, Sugai T. Development of a handpiece and probes for a microsurgical ultrasonic aspirator: instrumentation and application. Neurosurgery. 1999;45(5):1192–6; discussion 7.
26. Ridderheim PA, von Essen C, Zetterlund B. Indirect injury to cranial nerves after surgery with Cavitron ultrasonic surgical aspirator (CUSA). Case report. Acta Neurochir. 1987;89(1–2):84–6.
27. Epstein F. The Cavitron ultrasonic aspirator in tumor surgery. Clin Neurosurg. 1983;31:497–505.
28. Kanzaki J, Inoue Y, Kurashima K, Shiobara R. Use of the ultrasonically activated scalpel in acoustic neuroma surgery: preliminary report. Skull Base Surg. 2000;10(2):71–4.
29. Golub JS, Weber JD, Leach JL, Pottschmidt NR, Zuccarello M, Pensak ML, et al. Feasibility of the ultrasonic bone aspirator in retrosigmoid vestibular schwannoma removal. Otolaryngol Head Neck Surg. 2015;153(3):427–32.
30. Weber JD, Samy RN, Nahata A, Zuccarello M, Pensak ML, Golub JS. Reduction of bone dust with ultrasonic bone aspiration: implications for retrosigmoid vestibular schwannoma removal. Otolaryngol Head Neck Surg. 2015;152(6):1102–7.
31. Modest MC, Carlson ML, Link MJ, Driscoll CL. Ultrasonic bone aspirator (Sonopet) for meatal bone removal during retrosigmoid craniotomy for vestibular schwannoma. Laryngoscope. 2017;127(4):805–8.
32. Ansari SF, Terry C, Cohen-Gadol AA. Surgery for vestibular schwannomas: a systematic review of complications by approach. Neurosurg Focus. 2012;33(3):E14.
33. Teo MK, Eljamel MS. Role of craniotomy repair in reducing postoperative headaches after a retrosigmoid approach. Neurosurgery. 2010;67(5):1286–91; discussion 91–2.
34. Ito T, Mochizuki H, Watanabe T, Kubota T, Furukawa T, Koike T, et al. Safety of ultrasonic bone curette in ear surgery by measuring skull bone vibrations. Otol Neurotol. 2014;35(4):e135–9.
35. Levo H, Pyykkö I, Blomstedt G. Postoperative headache after surgery for vestibular schwannoma. Ann Otol Rhinol Laryngol. 2000;109(9):853–8.

Techniques of Dural Closure for Zero CSF Leak

12

Luciano Mastronardi, Guglielmo Cacciotti, Alberto Campione, Ali Zomorodi, Raffaelino Roperto, and Takanori Fukushima

Postoperative cerebrospinal fluid (CSF) leakage is a challenging and potentially hazardous condition that may complicate many complex cranial procedures. This is especially true for surgical approaches to the posterior cranial fossa, wherein a watertight dural reconstruction is not always feasible and CSF pulsation waves are greater than in other cranial areas [1–3]. Copeland et al. [4] reported that obesity, translabyrinthine (TL) approach, and longer operative times seem to significantly increase the risk of a CSF leak following vestibular schwannoma (VS) surgery. CSF fistulas into the soft tissues of the skull base can cause wound breakdown and/or pseudomeningocele, which often becomes very painful and debilitating. In addition, drainage of spinal fluid from the skin increases the risk for surgical site infections and meningitis [1]. In a series of 357 VSs, Nonaka et al. [3] reported a CSF leak in 7.6% of cases, wound infection in 2.2%, and meningitis in 1.7%. On the other hand, in a large systematic review of literature, Xia et al. [5] reported CSF leak complications in 1.6% (0.7–2.5%) of patients operated on for trigeminal neuralgia with microvascular decompression via retrosigmoid (RS) approach.

Diverse techniques for posterior cranial fossa dural reconstruction and closure have been reported: synthetic dural patches applied with continuous or separate stitches; incorporation of autologous tissues (pericranium or fascia lata); and augmentation with "muscle plugs" for small defects in the suture line and/or with gelatin sponge, absorbable hemostats, and dural sealants. Temporary CSF diversion can be employed via a lumbar drain or external ventricular drain to reduce the pressure gradient across the dural closure until complete wound healing [2]. Even using

L. Mastronardi (✉) · G. Cacciotti · A. Campione · R. Roperto
Department of Neurosurgery, San Filippo Neri Hospital—ASLRoma1, Rome, Italy
e-mail: mastro@tin.it

A. Zomorodi · T. Fukushima
Division of Neurosurgery, Duke University Medical Center, Carolina Neuroscience Institute, Raleigh, NC, USA
e-mail: ali.zomorodi@duke.edu; Fukushima@carolinaneuroscience.com

© Springer Nature Switzerland AG 2019
L. Mastronardi et al. (eds.), *Advances in Vestibular Schwannoma Microneurosurgery*, https://doi.org/10.1007/978-3-030-03167-1_12

these techniques, however, it is impossible to ensure a watertight dural closure for several reasons, including the gaps in the dura due to surgical needle during suturing. "Onlay" applicable synthetic dural grafts should be avoided because of the high hydrodynamic pressure of CSF in this district [2].

Chauvet et al. [6] developed an experimental device capable of testing dural closure watertightness. In their study, interrupted stitches sutures proved to have the same efficacy of running simple closures. In addition, the two sealants/glues (BioGlue®, CryoLife, USA, and DuraSeal®, Covidien, Ireland) and the two hemostatics (TachoSil®, Takeda, Japan, and Tissucol®, Baxter, USA) were tested and showed different watertightness properties [6]. Both the sealants significantly increased the watertightness of sutures; however, one sealant (DuraSeal®) and one hemostatic (TachoSil®) apparently yielded better results [6]. The nonaerosolized application of a thin layer of dural sealant (DuraSeal®) to the dry dural surface was also reported by Lam and Kasper [2]; in their study, the craniectomy defect was eventually repaired with a titanium mesh secured to the calvarium with microscrews [2].

The so-called surgical patch (TachoSil®, Takeda, Japan) combines the bioactive mechanism of action of fibrinogen and thrombin with the mechanical support of a collagen patch. It is derived from collagen—and therefore naturally resorbable— and is approved for hemostasis and tissue sealing. Upon contact with blood or other body fluids, the coagulation factors react to form a fibrin clot that sticks the surgical patch to the tissue surface, producing an air- and liquid-tight seal in few minutes and providing protection against postoperative rebleedings and leaks [6, 7].

Even if a variety of dural substitutes is currently available, many articles in the literature suggest that autologous materials are preferable compared to nonautologous substitutes [2, 8–10]. Czorny [8] used a pericranial graft harvested from the interparietal area to obtain tight dural closure for occipital craniotomy; this prevented pseudomeningocele and allowed for a better tolerance of possible postoperative cerebellar edema. Kosnik [9] proposed the technique of harvesting the ligamentum nuchae for dural closure after posterior fossa surgery: with this technique, the author avoided postoperative CSF leakage in more than 200 procedures. A vascularized pericranial flap or an autologous pericranium graft with dural sealant augmentation proved to be an effective way to repair the durotomy in posterior cranial fossa surgeries [2].

Autologous tissues for posterior cranial fossa dural closure have been mentioned in several articles [2, 8, 9]. Mastronardi et al. [11] reported the use of autologous pericranium harvested during the opening step of RS approach, inserted and stitched as an underlay hourglass-shaped plug for dural closure. Twenty-seven consecutive patients were enrolled in the study. At the end of the intracranial step of the procedure and after meticulous hemostasis, the autologous pericranium graft was inserted through the defect as an underlay hourglass-shaped plug. For obtaining this, the graft had to be slightly larger than the dural defect in order to have its edges under the dural plane. It was fixed to the dura mater under operative microscope magnification with separated stitches (with an "inside-to-outside" direction), using a 3-0 running silk Fig. 12.1. Afterward, the inserted patch was augmented with one layer

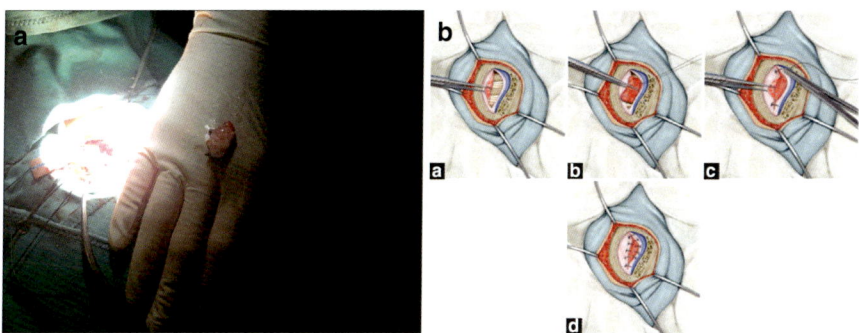

Fig. 12.1 Underlay hourglass-shaped autologous pericranium duraplasty in RS approach. (**a**) Autologous harvested pericranium. (**b**) Intraoperative picture of a typical RS approach closure. Step by step procedure is show: the local harvested pericranium (1) is inserted under the dural plane (2) with "inside-to-outside" direction (3) and is stitched (4). *Reprinted from Surgical Neurology International, 7:25, Luciano Mastronardi, Guglielmo Cacciotti, Franco Caputi, Raffaelino Roperto, Maria Pia Tonelli, Ettore Carpineta, Takanori Fukushima, Underlay hourglass-shaped autologous pericranium duraplasty in "key-hole" retrosigmoid approach surgery: Technical report, 2016, from Medknow under Creative Commons BY copyright license*

of absorbable hemostats (Fibrillary Surgicel, Ethicon, J and J, Somerville, New Jersey, USA), with small pieces of surgical patch (TachoSil®, Takeda, Japan), and with a dural sealant (DuraSeal, Covidien LLC, Mansfield, Massachusetts or Tisseel, Baxter, Deerfield, Illinois, USA). The authors did not observe any surgical site infections, meningitis, CSF leaks, or new neurological symptoms due to the dural closure technique. One (4% of the series) neurofibromatosis type 2 patient operated on for a large VS developed an asymptomatic small pseudomeningocele diagnosed on the 48-h postoperative CT scan; the CSF sac disappeared at the 3-month MRI follow-up [11].

References

1. Dubey A, Sung WS, Shaya M, Patwardhan R, Willis B, Smith D, et al. Complications of posterior cranial fossa surgery—an institutional experience of 500 patients. Surg Neurol. 2009;72(4):369–75.
2. Lam FC, Kasper E. Augmented autologous pericranium duraplasty in 100 posterior fossa surgeries—a retrospective case series. Neurosurgery. 2012;71(2 Suppl Operative):ons302–7.
3. Nonaka Y, Fukushima T, Watanabe K, Friedman AH, Sampson JH, Mcelveen JT, et al. Contemporary surgical management of vestibular schwannomas: analysis of complications and lessons learned over the past decade. Neurosurgery. 2013;72(2 Suppl Operative):ons103–15; discussion ons15.
4. Copeland WR, Mallory GW, Neff BA, Driscoll CL, Link MJ. Are there modifiable risk factors to prevent a cerebrospinal fluid leak following vestibular schwannoma surgery? J Neurosurg. 2015;122(2):312–6.
5. Xia L, Zhong J, Zhu J, Wang YN, Dou NN, Liu MX, et al. Effectiveness and safety of microvascular decompression surgery for treatment of trigeminal neuralgia: a systematic review. J Craniofac Surg. 2014;25(4):1413–7.

6. Chauvet D, Tran V, Mutlu G, George B, Allain JM. Study of dural suture watertightness: an in vitro comparison of different sealants. Acta Neurochir. 2011;153(12):2465–72.
7. Colombo GL, Bettoni D, Di Matteo S, Grumi C, Molon C, Spinelli D, et al. Economic and outcomes consequences of TachoSil®: a systematic review. Vasc Health Risk Manag. 2014;10:569–75.
8. Czorny A. Postoperative dural tightness. Value of suturing of the pericranium in surgery of the posterior cranial fossa. Neurochirurgie. 1992;38(3):188–90; discussion 90–1.
9. Kosnik EJ. Use of ligamentum nuchae graft for dural closure in posterior fossa surgery. Technical note. J Neurosurg. 1998;89(1):155–6.
10. Sameshima T, Mastronardi L, Friedman AH, Fukushima T. Microanatomy and dissection of temporal bone for surgery of acoustic neuroma and petroclival meningioma. 2nd ed. Raleigh, NC: AF Neurovideo; 2007.
11. Mastronardi L, Cacciotti G, Caputi F, Roperto R, Tonelli MP, Carpineta E, et al. Underlay hourglass-shaped autologous pericranium duraplasty in "key-hole" retrosigmoid approach surgery: technical report. Surg Neurol Int. 2016;7:25.

Face-to-Face Two-Surgeons Four-Hands Microsurgery

13

Takanori Fukushima and Ali Zomorodi

In the late 1960s and 1970s, Yasargil established micro-neurosurgery in the style of a one-man operation. This one-operator microsurgery became the gold standard from the 1980s to the millennium. Over the past 10 years, Fukushima has changed this "one man" show to two-surgeons four-hands micro-neurosurgery using face-to-face ocular arrangement of the modern floating operating microscope. In the traditional microsurgical setup, there is very limited access to the surgical field by anyone other than the operating surgeon. In conventional microscope surgery, the assistant stands to the surgeon's right or left side using the 2-D observer side scope. It has been very difficult for neurosurgical trainees and younger faculties to learn actual microsurgical dissection techniques just by observing an operation by an expert. Over the past decade, Fukushima adopted a new microsurgical method using a face-to-face microscope arrangement. The modern microscope provides face-to-face ocular arrangement in which two surgeons observe equally with three-dimensional viewing field. This two-surgeons four-hands microsurgery will facilitate any microsurgical procedure from cerebrovascular to neoplastic microsurgical events. In particular, four-hands surgery will facilitate the microanastomosis bypass procedures, such as co-surgeon performs suction, holding, and efficiently helping the principal surgeon's micromaneuvers. Any surgical procedure around the neck, carotid artery, or spine, surgery will be facilitated in both the microsurgical maneuver and time-consuming efficiencies with shorter operating time. The scrub assistant can be by the right side or left side of the principal surgeon or on the right side of the co-surgeon position, such as Fig. 13.1 illustrates. Figure 13.2 illustrates the photograph of Fukushima and Zomorodi operating face-to-face. This new strategy of two surgeons requires understanding of the alternative surgical anatomy shifted

T. Fukushima (✉) · A. Zomorodi
Division of Neurosurgery, Duke University Medical Center, Carolina Neuroscience Institute, Raleigh, NC, USA
e-mail: Fukushima@carolinaneuroscience.com; ali.zomorodi@duke.edu

© Springer Nature Switzerland AG 2019
L. Mastronardi et al. (eds.), *Advances in Vestibular Schwannoma Microneurosurgery*, https://doi.org/10.1007/978-3-030-03167-1_13

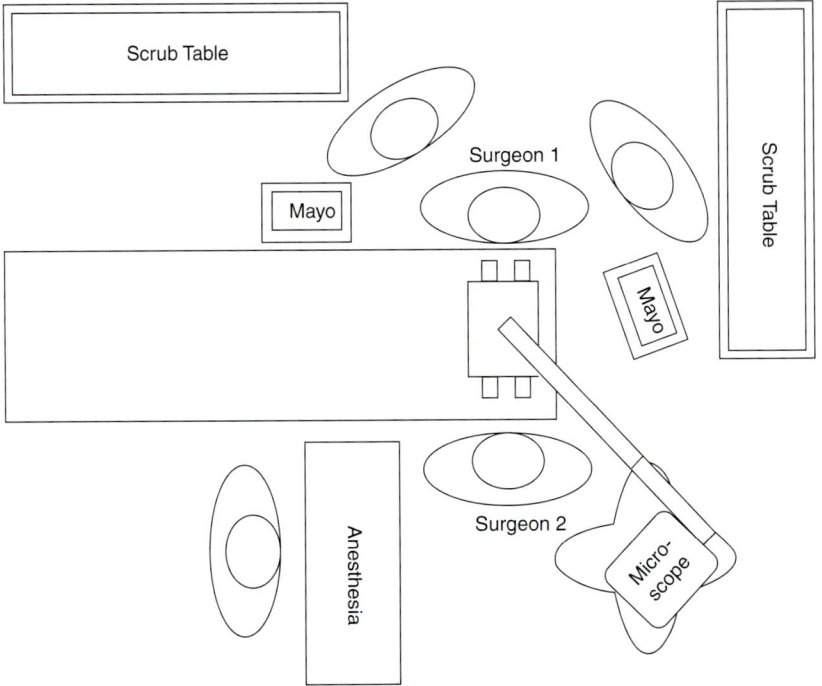

Fig. 13.1 Arrangement in OR

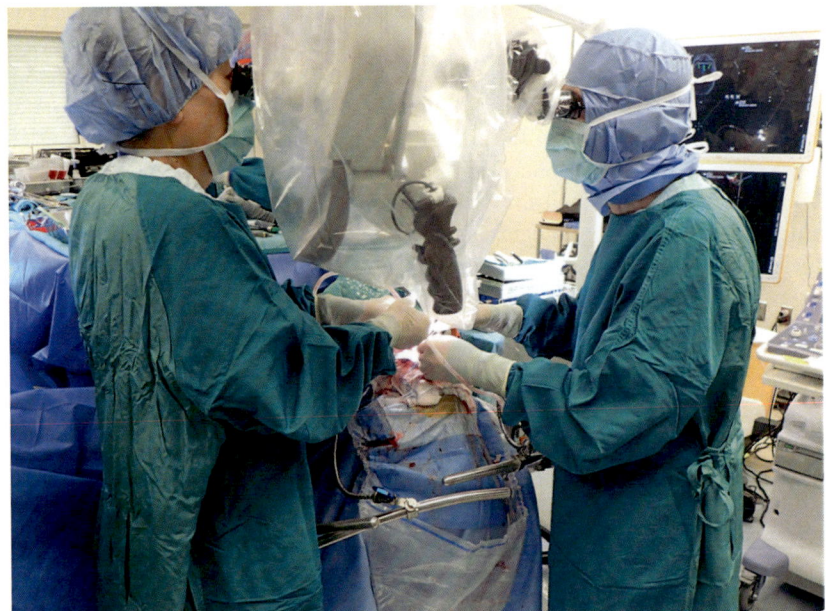

Fig. 13.2 Two men four hands

90°–180°, relative to the standard looking view by a single surgeon. Neurosurgeons using this face-to-face microscope arrangement must learn the 90° or 180° shifted anatomical details with preoperative practice and exercises. In this two-surgeons four-hands style of microsurgery, Fukushima designed wide dynamic range universal holding system that provides multiple blunt hooks for fixing the soft tissues, drip irrigation, flexible brain spatula holder, and patty board. This system is extremely useful to effectively increase the number of artificial arms available for surgical holding. As compared to the previous Fukushima simple retractor holder published in 1980 [1] or Sugita craniotomy frame published in 1978 [2] or other similar systems such as Greenberg or Budde Halo frames, the conventional retractor holder has limited use with very busy operative field with short arms, bars, and brain spatula. The wide range Universal Holding System (as shown in Fig. 13.3) allows wide access to the operative field while holding soft tissues and the brain surface without disturbing the surgeon's operative field. The Universal Holding System is available from many surgical instrument corporations in the USA and Japan. This face-to-face technique has been especially helpful for the treatment of complex vascular lesions, bypass procedures, and vascular skull base tumors. Face-to-face two-surgeons four-hands method can facilitate the setup for the use of double suctions, double dissectors, double microscissors, and double bipolar system. The operative area is covered by the four vertical posts and six curved bars. The surgeon's arms can be used for arm rest or hand rest. The side bars and front bars will hold multiple soft tissue blunt hooks, flexible holders, patty board, and particularly the continuous drip irrigation blunt needle, which had been published in 1993 by Day and Fukushima [3]. The two-operator style provides very useful educational purposes. The resident or assistant surgeon is directly involved in the operation with real-time hands-on education. To explain practical clinical microsurgical anatomy and to teach sharp dissection and clean separation of the arachnoid membrane and gentle dissection of arteries and veins and cranial nerves, younger neurosurgeons will learn quickly and very accurately the senior surgeon's micro-operative techniques and maneuvers. Particularly in acoustic neuroma surgery, as the patient is positioned lateral, the main surgeon operates from the occiput side, and the assisting surgeon will be in the front position. Therefore, both surgeons must learn 180° shifted microanatomy of the cerebellopontine angle. In recent years, several reports have claimed that they use no brain retraction and no brain spatula and stressing or emphasizing that "retractorless" surgery is superior to the conventional traditional way of utilizing the tapered brain spatula. My experience for any brain surgery shows that the 2 mm tapered brain spatula is extremely helpful and important to hold the brain preventing the brain from sagging down and gently allowing micro-dissection with two free hands [4]. Publications claiming retractorless surgery do retract the brain and nerves with suckers and instruments. It is certain that such surgery doing continuous or intermittent brain retraction with suckers and micro instruments would unwittingly cause more brain injury; therefore, Fukushima strongly recommends the style of two-surgeons four-hands micro-neurosurgery and the use of a universal holding system.

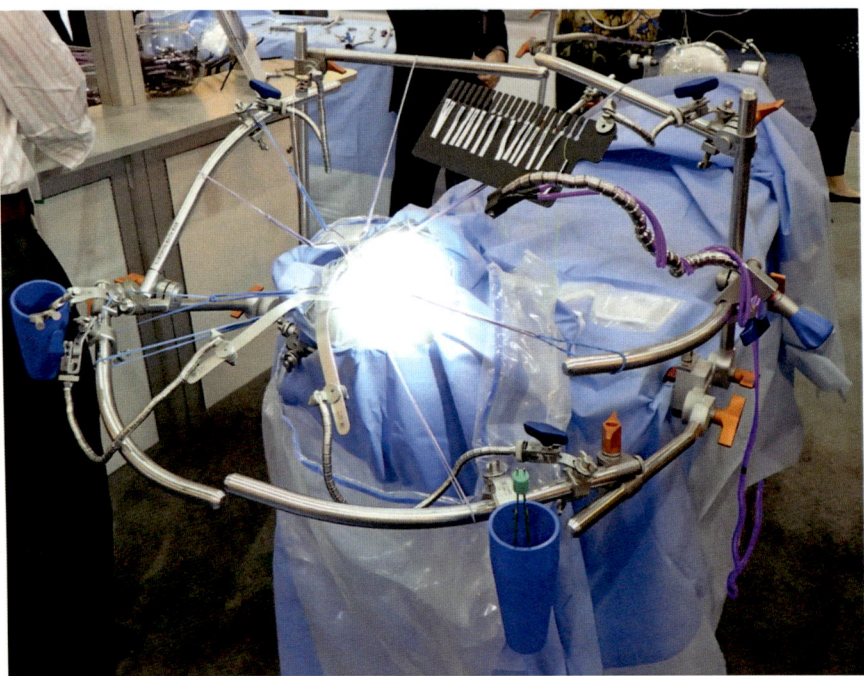

Fig. 13.3 TSI: Fukushima head frame

References

1. Fukushima T, Sano K. Simple retractor holder for the Mayfield skull clamp. Surg Neurol. 1980;13:320.
2. Sugita K, Hirota T, Mizutani T, et al. A newly designed multipurpose microneurosurgical head frame. J Neurosurg. 1978;48:656–7.
3. Day JD, Fukushima T. Two simple devices for microneurosurgery: automatic drip irrigating needle and a suction retractor. Neurosurgery. 1993;32:867–8.
4. Zomorodi A, Fukushima T. Two surgeons 4 hands micro neurosurgery with universal holder system: Technical note. Neurosurg Rev. 2017;40:523–6.

Part IV

Projects in Progress

Diluted Papaverine for Microvascular Protection of Cranial Nerves

14

Alberto Campione, Carlo Giacobbo Scavo,
Guglielmo Cacciotti, Raffaelino Roperto,
and Luciano Mastronardi

14.1 Pathogenesis of Vasospasm After Cranial Case Tumor Resection

Cerebral vasospasm is well known to occur after various cerebral neurosurgical events that cause subarachnoid hemorrhage. However, cerebral vasospasm can occur after cranial base tumor resection [1]. Bejjani et al. [2] described a series of 470 consecutive patients with cranial base tumors; 9 (7 meningiomas, 1 chordoma, and 1 trigeminal schwannoma) cases of postoperative cerebral vasospasm (1.9% of the total population) were reported, and the authors reviewed the pathogenesis at the basis of such occurrences. In all of the cases, the tumors were located close to the basal cisterns, and marked intraoperative bleeding was reported in some of the surgical procedures, thus raising the suspicion that blood spillage into the basal cisterns could be a potential risk factor. Blood breakdown products are the most frequently evoked etiology of vasospasm after subarachnoid hemorrhage. The reported occurrence of vasospasm 8 days after the skull base tumor resection was similar to the interval seen after subarachnoid hemorrhage and may indicate a similar mechanism. Manipulation of the major vessels of the basal cisterns was addressed as another possible risk factor. A statistically significant difference was observed in the incidence of vessel encasement and vessel narrowing between patients with and without vasospasm. Direct mechanical irritation of the smooth muscle cells or the vasa nervorum was supposed to be the mechanism in this case [2].

A. Campione (✉) · C. Giacobbo Scavo · G. Cacciotti · R. Roperto · L. Mastronardi
Department of Neurosurgery, San Filippo Neri Hospital—ASLRoma1, Rome, Italy
e-mail: mastro@tin.it

© Springer Nature Switzerland AG 2019
L. Mastronardi et al. (eds.), *Advances in Vestibular Schwannoma Microneurosurgery*, https://doi.org/10.1007/978-3-030-03167-1_14

14.2   Vasospasm After VS Resection

Krayenbühl [3, 4] first reported a case of internal carotid spasm after VS resection in 1960. Another case of VS with a two-stage removal due to severe bleeding was presented by de Almeida et al. [5] in 1985. The authors reported that the cerebrospinal fluid (CSF) was bloody, vasospasm was shown in the angiograms, and an ischemic area was disclosed on the CT scan. The outcome and the neuroradiologic examinations suggested that blood in the basal cisterns caused the vasospasm and the brain ischemia [5], as Bejjani et al. [2] would later confirm as the principal cause of vascular complications in skull base surgery. In 2003, Kania et al. [6] reported six cases of vascular complications occurring after VS surgery: one case of cerebral vasospasm, three cases of arterial infarction in the territory of anterior inferior cerebellar artery (AICA), one case of hematoma in the cerebellopontine angle (CPA), and one last case of venous infarction of the cerebellar vermis. Traumatic injury to the vessels and/or thrombosis was addressed as the main causes of such occurrences [6].

In 2015, Qi et al. [7] conducted a study to identify factors associated with postoperative cerebral vasospasm in patients with VS. The VSs were removed using the retrosigmoid approach, with particular care to minimize bleeding and protect the facial, trigeminal, and lower cranial nerves and brainstem. Flow velocities in the bilateral internal carotid, middle cerebral, and anterior cerebral arteries, assessed with transcranial Doppler ultrasonography before and after surgery, were used to detect cerebral vasospasm. Forty-three (53.8%) of the 80 patients included were diagnosed with cerebral vasospasm [7]. The findings of this study suggested that more than half of the patients experienced asymptomatic vasospasm that was identifiable by transcranial Doppler ultrasonography; it should be emphasized that patients with abnormalities in postoperative vital signs, electrolytes, and blood gases were excluded from the present analysis. Younger patient age, larger tumor size, and firmer tumor consistency were independently associated with postoperative cerebral vasospasm after univariate and multivariate analysis [7]. The study by Qi et al. [7] demonstrates that silent vasospasm is much more common than previously thought among patients undergoing VS surgery; this might mean that the cerebral circulation shows a high reactivity to surgical trauma, and it can be hypothesized that this happens not only in elastic and muscular arteries but also in arterioles—and especially those undergoing surgical trauma within the CPA. Many studies have reported that hearing preservation failure may be due to traumatic surgical injury to the internal auditory artery (IAA); however, it cannot be excluded that the IAA undergoes vasospasm, which would be the logical consequence of its manipulation during surgery [8]. The mechanism by which mechanical manipulations may induce IAA vasospasm may involve arterial neuro-regulation. There is evidence that the cochlear blood flow (CBF) is at least partially under sympathetic nervous system control [9]. In vessels found in the subarachnoid space, external mechanical compression of arterial walls induced vasoconstriction by exciting sympathetic nerve fibers found within the surrounding arachnoid strands [9]. From this perspective, prevention of vasospasm in VS surgery is useful for microvascular protection of the cranial nerves and especially the cochlear nerve—data over vascular deficiency of the facial nerve is missing in the literature.

The only animal model existing in the literature of IAA vasospasm was proposed by Morawski et al. [9]. Vasospasm was mechanically induced by compressing the IAA in the control ears of six rabbits after application of topical saline. The subsequent reduction of CBF was measured using a laser-Doppler (LD) flow-monitoring technique. Functional loss of cochlear activity was verified with distortion product otoacoustic emissions (DPOAE). The contralateral experimental ears were treated with the topical application of papaverine directly to the IAA and cochleo-vestibular nerve complex. CBF and DPOAE were compared between the control and papaverine treated ears for 3-min and 5-min IAA compressions. Every control ear demonstrated some degree of post-compression IAA vasospasm (i.e., reduced CBF) and reduction of DPOAE. Nearly complete recovery of CBF and DPOAE to baseline was observed in all of the papaverine treated ears. The data from this study strongly suggested that topical application of papaverine before dissection in the internal auditory canal/CPA may prevent reduction of CBF and resulting cochlear outer hair cell (OHC) function by eliminating vasospasm of the IAA during human surgery [9].

14.3 Papaverine: Mechanism of Action and Routes of Administration

Papaverine is a benzylisoquinoline alkaloid derived from opium and is known to be a potent vasodilator directly acting on smooth muscle and causing it to relax. Its mechanism of action is thought to be inhibition of cyclic adenosine monophosphate (cAMP) and cyclic guanosine monophosphate (cGMP) phosphodiesterases in smooth muscle, leading to increased intracellular levels of cAMP and cGMP and eventually to nonspecific smooth muscle relaxation. Papaverine may also inhibit the release of calcium from the intracellular space by blocking calcium ion channels in the cell membrane [10, 11]. In addition to its spasmolytic effect, it inhibits collagen-induced platelet aggregation and serotonin release [12].

Intra-arterial administration of papaverine has been used to treat arterial vasospasm after aneurysmal subarachnoid hemorrhage [13]. Commercially available injectable solutions are typically quite acidic (pH 3–4.5) and can be caustic to the vascular endothelium; apoptosis of vascular endothelial and smooth muscle cells in animal investigations has been observed after papaverine injection [14]. When applied topically, it acts as a smooth muscle relaxant to break intraoperative vasospasm. Topical administration of papaverine has been used in two fields wherein microvascular protection and preservation is fundamental: plastic and reconstructive microsurgery [12, 15–18] and neurovascular surgery [19–21]. In the latter case, topical administration is necessarily intracisternal and constitutes a valid—although infrequent—alternative to the intra-arterial injection.

14.4 Topical Papaverine in Microsurgery and Neurovascular Surgery

Topical vasodilators are widely used in reconstructive microsurgery in order to ameliorate intraoperative vascular spasm (vasospasm) and facilitate microvascular anastomoses during free tissue transfer. As vasospasm has been reported after VS resection [5, 6] and studied on animal models in the same context [9], studies about topical papaverine administration in reconstructive microsurgery may serve as an indirect model for VS surgery, too. A survey of plastic surgeons in the United Kingdom conducted by Yu et al. in 2010 revealed that 94% of surgeons surveyed routinely used vasodilators intraoperatively; papaverine, verapamil, and lidocaine were the preferred agents [12]. However, justification for their use, agent of choice, and technique of administration varied widely due to the lack of systematic reviews in the literature [12]. Between 2014 and 2016, three reviews [16–18] were published concerning the evidence at the basis of topical vasodilator administration; all of the studies had limitations reflecting those of currently available literature, which is scarce and, in certain cases, obsolete with some of the studies more than 30 years old [18]. Vargas et al. [18] performed a systematic review of the literature to identify articles relevant to pharmacologic treatment of intraoperative vasospasm in vivo. As regards papaverine, it was found to improve arterial microanastomosis patency relative to saline [22] and to provide effective relief (blood flow 116% relative to control) and some degree of prevention of vasospasm. In addition, papaverine had a vasodilatory effect on nonspastic vessels, resulting in an absolute increase in flow [23]. Time to effect was reported to range from 1 to 5 min after topical application in several studies [15, 23–25]. Overall, papaverine was deemed an effective spasmolytic and antispasmodic topical agent with quick onset and a reasonable duration of effect [16, 18]. In their systematic review, Rinkinen and Halvorson [17] reported data confirming the characteristics of papaverine as previously described by Vargas et al. [18] and compared it to two other classes of topical vasodilators as local anesthetics and calcium channel blockers (CCB). Based on the studies examined, CCBs (nicardipine/nifedipine/verapamil) appeared to have the greatest efficacy in preventing vasospasm and inducing vasodilation following microsurgical anastomosis compared with other agents. CCBs, in general, were reported as more efficacious and versatile compared with papaverine and lidocaine with regard to sustained vasodilation and side effect profile [17]. On the other hand, the comparison between papaverine and lidocaine resulted in papaverine being superior to lidocaine (1%) in sustaining vasodilation in animal models such as in Kerschner and Futran (rat femoral artery model) [15]. It is interesting to observe how the research on vasospasm prevention in VS surgery has investigated CCBs in the last years, too. However, no direct correlations can be inferred in this case between reconstructive microsurgery and VS surgery. In fact, the study by Rinkinen and Halvorson was mainly concerned on outcomes of microvascular anastomoses and not vasospasm in general. In addition, the use of nimodipine in the treatment of VSs—according to the latest evidence-based guidelines by the Congress of Neurological Surgeons [Van Gompel]—has been recommended not as a topical agent but as an enteral/parenteral preoperative adjunct. Therefore, nimodipine still constitutes a very interesting emerging adjunct therapy for VSs, but further studies are warranted to compare its efficacy in preventing postoperative vascular complications with that of papaverine.

The use of papaverine in neurovascular surgery is mainly aimed at alleviating vasospasm as a complication of subarachnoid hemorrhage. Persistence of subarachnoid blood after VS resection may induce vasospasm [6] and pathophysiologically resemble the events following a subarachnoid hemorrhage; from this point of view, neurovascular research constitutes an indirect model of and an insight into what happens to the cerebellopontine angle (CPA) microvasculature during VS surgery. Pennings et al. [20] studied the microvascular responses to papaverine in patients undergoing aneurysm surgery to test the hypothesis that cerebral arterioles have a reduced capacity to dilate after subarachnoid hemorrhage. In 14 patients undergoing aneurysm surgery, the diameter changes of cortical microvessels after topical application of papaverine were observed using orthogonal polarizing spectral imaging. In control subjects, neither arterioles nor venules showed diameter changes; instead, in patients operated ≤ 48 h after subarachnoid hemorrhage, papaverine resulted in vasodilation of arterioles with $45 \pm 41\%$ increase in arteriolar diameter ($P < 0.012$) [20]. Therefore, cortical microvessels proved to be responsive to the topical vasodilator, and papaverine was shown to exert a significant vasodilating effect. Although indirectly, the results of this study may also be deemed valid for skull base and VS surgery: intraoperative bleeding supposedly inducing vasospasm in the CPA microvessels is likely to be relieved by topical papaverine administration prior to dural closure, which is indeed an early intervention on a controlled and confined sort of subarachnoid hemorrhage. Dalbasti et al. [19] proposed local application of papaverine in a biodegradable controlled—or sustained—release matrix for vasospasm prophylaxis in patients scheduled for aneurysm surgery. Controlled-release papaverine (papaCR) drug pellets were prepared using the biodegradable aliphatic polyester poly(DL-lactide-co-glycolide) as the carrier matrix. During aneurysm surgery, drug pellets were placed in cisterns over arterial segments. No adverse effects due to the drug were observed. The PapaCR effectively prevented development of clinical vasospasm and, as far as outcome scores were concerned, average Glasgow Outcome Scale scores were 4.93 ± 0.05 in the PapaCR-treated group and 3.84 ± 1.63 in the control group [19]. Praeger et al. [21] first described sustained reversal of severe symptomatic vasospasm by intraoperative topical papaverine in combination with aneurysm repair in a SAH patient. In fact, papaverine is usually given therapeutically at angiography, and previous reports described instillation into the surgical bed only for prophylaxis of vasospasm [19, 26]. The authors deemed the procedure generally safe and suggested that it should be considered as a treatment option for similar patients, who present with unsecured aneurysms not amenable to endovascular therapy in the setting of severe vasospasm [21]. Put into the perspective of VS surgery, the results of this study give a pathophysiological justification for topical papaverine administration at the end of the intradural phase of the procedure in order to relieve the CPA microvasculature vasospasm.

14.5 Papaverine and Cranial Nerves: Preservation or Damage?

Postoperative neurological side effects of topical (or intracisternal) papaverine include transient cranial nerve palsies, most commonly mydriasis due to oculomotor nerve involvement with rapid resolution [27–29]. Papaverine toxicity is believed to occur in the setting of its antimuscarinic action, and blood-CSF and blood-brain barrier compromise owing to subarachnoid hemorrhage (or similar events) and papaverine direct effect [28].

The most important side effects of topical papaverine in VS surgery are those reported on facial and cochlear nerve. Lang et al. [26] reported a case of transient mydriasis and prolonged facial nerve palsy after intracisternal papaverine application during elective clipping of an unruptured middle cerebral artery aneurysm; the facial dysfunction persisted for 2 months before complete recovery. The authors hypothesized that prolonged irrigation of the cisterns could have washed the papaverine into contact with the facial nerve [26]. Liu et al. [30] reported a case of transient facial nerve palsy that occurred after papaverine was topically applied during a hearing preservation VS removal. During tumor removal, a solution of 3% papaverine soaked in a Gelfoam pledget was placed over the cochlear nerve. Shortly thereafter, the quality of the facial nerve stimulation deteriorated markedly. Electrical stimulation of the facial nerve did not elicit a response at the level of the brainstem but was observed to elicit a robust response more peripherally. Immediately after surgery, the patient had a House-Brackmann (HB) grade 5 facial palsy [31]. After several hours, this improved to a HB 1. At the 1-month follow-up examination, the patient exhibited normal facial nerve function and stable hearing. The authors concluded that intracisternal papaverine may cause a transient facial nerve palsy by producing a temporary conduction block of the facial nerve [30].

Chadwick et al. [32] conducted a retrospective review of 11 microvascular decompression operations wherein topical papaverine was used as a direct therapeutic action to manage vasospasm. A temporal relationship was found between topical papaverine and adverse brainstem evoked auditory potentials (BAEP) changes leading to complete waveform loss. The average temporal delay between papaverine and the onset of BAEP change was 5 min. In 10 of 11 patients, BAEP waves II/III–V completely disappeared within 2–25 min after papaverine. One patient showed no recovery of later waves and a delayed profound sensorineural hearing loss. The average recovery time of BAEP waveforms to pre-papaverine baseline values was 39 min. The complete disappearance of BAEP waveforms with a consistent temporal delay suggested a possible adverse effect on the proximal cochlear nerve. The authors recommended keeping the papaverine away from the proximal cochlear nerve to avoid complications in auditory function. Dilution of papaverine in saline prior to application was recommended. To control the spread of papaverine, the authors also suggested to place a small papaverine-soaked Gelfoam pledget against the spastic arterial segment until the spasm resolved. Zhou et al. [28] suggested that intracisternal papaverine at a concentration of 0.3% would reasonably diminish the risk for neurotoxicity [28].

References

1. Aoki N, Origitano TC, al-Mefty O. Vasospasm after resection of skull base tumors. Acta Neurochir. 1995;132(1–3):53–8.
2. Bejjani GK, Sekhar LN, Yost AM, Bank WO, Wright DC. Vasospasm after cranial base tumor resection: pathogenesis, diagnosis, and therapy. Surg Neurol. 1999;52(6):577–83; discussion 83–4.
3. Krayenbuhl H. [Not available]. Schweiz Med Wochenschr 1959;89(8):191–5.
4. Krayenbühl H. Beitrag zur Frage des cerebralen angiopastischen Insults. Schweiz Med Wochenschr. 1960;90:961–5.
5. de Almeida GM, Bianco E, Souza AS. Vasospasm after acoustic neuroma removal. Surg Neurol. 1985;23(1):38–40.
6. Kania R, Lot G, Herman P, Tran Ba Huy P. [Vascular complications after acoustic neurinoma surgery]. Ann Otolaryngol Chir Cervicofac 2003;120(2):94–102.
7. Qi J, Jia W, Zhang L, Zhang J, Wu Z. Risk factors for postoperative cerebral vasospasm after surgical resection of acoustic neuroma. World Neurosurg. 2015;84(6):1686–90.
8. Mom T, Montalban A, Khalil T, Gabrillargues J, Chazal J, Gilain L, et al. Vasospasm of labyrinthine artery in cerebellopontine angle surgery: evidence brought by distortion-product otoacoustic emissions. Eur Arch Otorhinolaryngol. 2014;271(10):2627–35.
9. Morawski K, Telischi FF, Merchant F, Namyslowski G, Lisowska G, Lonsbury-Martin BL. Preventing internal auditory artery vasospasm using topical papaverine: an animal study. Otol Neurotol. 2003;24(6):918–26.
10. Cooper GJ, Wilkinson GA, Angelini GD. Overcoming perioperative spasm of the internal mammary artery: which is the best vasodilator? J Thorac Cardiovasc Surg. 1992;104(2):465–8.
11. Newell DW, Elliott JP, Eskridge JM, Winn HR. Endovascular therapy for aneurysmal vasospasm. Crit Care Clin. 1999;15(4):685–99, v.
12. Yu JT, Patel AJ, Malata CM. The use of topical vasodilators in microvascular surgery. J Plast Reconstr Aesthet Surg. 2011;64(2):226–8.
13. Kassell NF, Helm G, Simmons N, Phillips CD, Cail WS. Treatment of cerebral vasospasm with intra-arterial papaverine. J Neurosurg. 1992;77(6):848–52.
14. Gao YJ, Stead S, Lee RM. Papaverine induces apoptosis in vascular endothelial and smooth muscle cells. Life Sci. 2002;70(22):2675–85.
15. Kerschner JE, Futran ND. The effect of topical vasodilating agents on microvascular vessel diameter in the rat model. Laryngoscope. 1996;106(11):1429–33.
16. Ricci JA, Koolen PG, Shah J, Tobias AM, Lee BT, Lin SJ. Comparing the outcomes of different agents to treat vasospasm at microsurgical anastomosis during the papaverine shortage. Plast Reconstr Surg. 2016;138(3):401e–8e.
17. Rinkinen J, Halvorson EG. Topical vasodilators in microsurgery: what is the evidence? J Reconstr Microsurg. 2017;33(1):1–7.
18. Vargas CR, Iorio ML, Lee BT. A systematic review of topical vasodilators for the treatment of intraoperative vasospasm in reconstructive microsurgery. Plast Reconstr Surg. 2015;136(2):411–22.
19. Dalbasti T, Karabiyikoglu M, Ozdamar N, Oktar N, Cagli S. Efficacy of controlled-release papaverine pellets in preventing symptomatic cerebral vasospasm. J Neurosurg. 2001;95(1):44–50.
20. Pennings FA, Albrecht KW, Muizelaar JP, Schuurman PR, Bouma GJ. Abnormal responses of the human cerebral microcirculation to papaverin during aneurysm surgery. Stroke. 2009;40(1):317–20.
21. Praeger AJ, Lewis PM, Hwang PY. Topical papaverine as rescue therapy for vasospasm complicated by unsecured aneurysm. Ann Acad Med Singap. 2014;43(1):62–3.
22. Swartz WM, Brink RR, Buncke HJ. Prevention of thrombosis in arterial and venous microanastomoses by using topical agents. Plast Reconstr Surg. 1976;58(4):478–81.
23. Hou SM, Seaber AV, Urbaniak JR. Relief of blood-induced arterial vasospasm by pharmacologic solutions. J Reconstr Microsurg. 1987;3(2):147–51.

24. Evans GR, Gherardini G, Gürlek A, Langstein H, Joly GA, Cromeens DM, et al. Drug-induced vasodilation in an in vitro and in vivo study: the effects of nicardipine, papaverine, and lidocaine on the rabbit carotid artery. Plast Reconstr Surg. 1997;100(6):1475–81.
25. Gherardini G, G,rlek A, Cromeens D, Joly GA, Wang BG, Evans GR. Drug-induced vasodilation: in vitro and in vivo study on the effects of lidocaine and papaverine on rabbit carotid artery. Microsurgery. 1998;18(2):90–6.
26. Lang EW, Neugebauer M, Ng K, Fung V, Clouston P, Dorsch NW. Facial nerve palsy after intracisternal papaverine application during aneurysm surgery—case report. Neurol Med Chir (Tokyo). 2002;42(12):565–7.
27. Zhou W, Ma C, Huang C, Yan Z. Intra- and post-operational changes in pupils induced by local application of cisternal papaverine during cerebral aneurysm operations. Turk Neurosurg. 2014;24(5):710–2.
28. Zhou X, Alambyan V, Ostergard T, Pace J, Kohen M, Manjila S, et al. Prolonged intracisternal papaverine toxicity: index case description and proposed mechanism of action. World Neurosurg. 2018;109:251–7.
29. Zygourakis CC, Vasudeva V, Lai PM, Kim AH, Wang H, Du R. Transient pupillary dilation following local papaverine application in intracranial aneurysm surgery. J Clin Neurosci. 2015;22(4):676–9.
30. Liu JK, Sayama CM, Shelton C, MacDonald JD. Transient facial nerve palsy after topical papaverine application during vestibular schwannoma surgery. Case report. J Neurosurg. 2007;107(5):1039–42.
31. House JW, Brackmann DE. Facial nerve grading system. Otolaryngol Head Neck Surg. 1985;93(2):146–7.
32. Chadwick GM, Asher AL, Van Der Veer CA, Pollard RJ. Adverse effects of topical papaverine on auditory nerve function. Acta Neurochir. 2008;150(9):901–9; discussion 9.

Flexible Endoscope for IAC Control of Tumor Removal

15

Alberto Campione, Carlo Giacobbo Scavo, Guglielmo Cacciotti, Raffaelino Roperto, and Luciano Mastronardi

Intracanalicular vestibular schwannomas (ICVS) were defined as those vestibular schwannomas (VS) that are limited to the internal auditory canal (IAC) without extension into the cerebellopontine angle (CPA) [1]. Their incidence has progressively increased thanks to the widespread accessibility to magnetic resonance imaging (MRI) scans and accounts for around 8% of all VSs [2]. The best treatment modality of ICVS is still a matter of debate, and the most common options besides surgery are "wait and see" or radiosurgical treatment [3]. However, the natural course of auditory function in patients harboring purely ICVSs showed that a mild but progressive hearing loss has to be expected [4], particularly in the first years after diagnosis [2]. According to outstanding experienced neurosurgeons [5], microsurgery should be considered as the first treatment option in patients suitable for surgery with documented AAO-HNS class A–B of hearing [6]. In fact, when the patient has a good preoperative hearing function, the aim of surgery is to achieve complete tumor removal with preservation of hearing.

Thus, surgical treatment appears to be an excellent option as it allows for attempt of hearing preservation, high rate of facial nerve preservation, and improvement of vestibular function after surgery [7]. However, surgical treatment can be challenging, with potential risks for surgical morbidity, since the majority of patients has a good clinical status. Hearing preservation also implies anatomical respect of the inner ear structures. Using a retrosigmoid approach, the internal auditory meatus cannot be completely opened to expose the fundus in order to preserve the superior and posterior semicircular canals. Therefore, a straight microscopic view cannot provide an adequate visualization of the most lateral part of IAC, forcing the surgeon to work blind with hooklets and curettes around the meatal bone. A possible solution to this limit was advocated by Mazzoni et al. [8] who recently described the surgical technique of microsurgical retrolabyrinthine meatotomy for reaching the fundus of IAC

A. Campione (✉) · C. Giacobbo Scavo · G. Cacciotti · R. Roperto · L. Mastronardi
Department of Neurosurgery, San Filippo Neri Hospital—ASLRoma1, Rome, Italy
e-mail: mastro@tin.it

© Springer Nature Switzerland AG 2019
L. Mastronardi et al. (eds.), *Advances in Vestibular Schwannoma Microneurosurgery*, https://doi.org/10.1007/978-3-030-03167-1_15

by means of a careful exposure of the entrance to the Fallopian canal [8]. The same approach, i.e., opening the intrameatal canal with labyrinth preservation, was described in a cadaveric study by Pillai and colleagues confirming an excellent exposure of the fallopian portion of the fundus of the IAC [9]. However, even if an extensive drilling of the posterior wall of the IAC is performed, the vestibular portion of the fundus cannot be adequately exposed if preservation of labyrinthine structures is intended, as they are crucial to obtain hearing preservation. Nonetheless, the vestibular quadrant is the anatomical region of the IAC where VSs usually arise from the inferior or superior vestibular nerves. Therefore, residual tumor can be frequently encountered in this area, strongly attached to the fundus of IAC.

Endoscope assistance provides optimal views toward the fundus of the IAC allowing access to the most lateral part of the tumor under visual control [10–12]. Therefore, the use of the endoscope decreases the amount of bone drilling in the posterior wall of the IAC and the risk of injury to the superior and posterior semicircular canals. Moreover, as demonstrated in a recent paper by Abolfotoh et al. [13], endoscopic assistance also improves the ability to evaluate the extent of resection intraoperatively (Fig. 15.1). Indeed, the exclusive utilization of

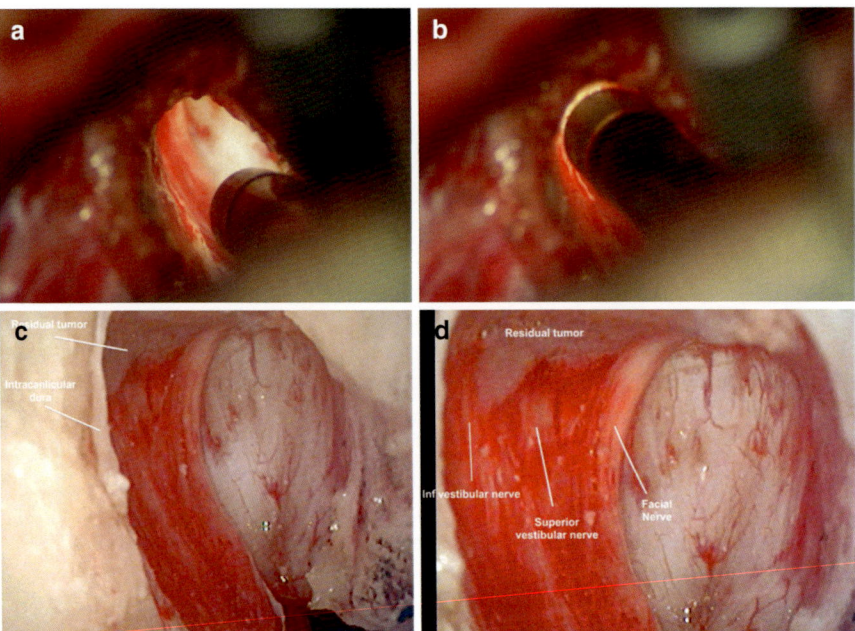

Fig. 15.1 Flexible endoscope assistance. (**a**, **b**) The tip of the flexible endoscope is progressively introduced into the IAC so that the fundus can be visualized. (**c**) Endoscopic view of the IAC as the flexible endoscope is positioned as illustrated in (**a**). (**d**) Endoscopic view of the IAC as the flexible endoscope is fully inserted into the canal, as shown in (**b**). The facial nerve is visible along with the vestibular nerves and the tumor remnant. *Reprinted from World Neurosurgery, 115, Francesco Corrivetti, Guglielmo Cacciotti, Carlo Giacobbo Scavo, Raffaelino Roperto, Luciano Mastronardi, Flexible Endoscopic-Assisted Microsurgical Radical Resection Of Intracanalicular Vestibular Schwannomas By Retrosigmoid Approach: Operative Technique, Pages No. 229–233, 2018, with permission from Elsevier*

microscopic view has a poor reliability in evaluating intraoperatively the extent of resection of CPA tumors with deep IAC extension [13]. Furthermore, the endoscopic visualization of the fundus of IAC enables the surgeon to detect the exact position of the residual tumor into the IAC, thus guiding the microsurgical dissection and allowing for complete tumor resection (Fig. 15.1).

Endoscopic-assisted techniques in the surgical treatment of CPA tumors have been well studied for decades, and excellent surgical results in achieving additional safe tumor resection of intracanalicular lesions have recently been described in the literature [13–16]. However, only one recent article described the utility of endoscopic technique in the surgical management of ICVSs [17]. Corrivetti et al. [18] reported the surgical treatment of three cases of ICVSs operated on with an endoscopic-assisted microneurosurgical retrosigmoid approach by means of a flexible endoscope (4-mm × 65-cm, Karl Storz, Inc.).

The three cases were initially diagnosed on the basis of vestibular dysfunction (rotational vertigo and dizziness) as presenting symptoms. The duration of the symptoms ranged from 2 to 24 months. All patients had AAO-HNS class A–B of hearing at initial presentation, corresponding to both a pure tone audiometry (PTA) threshold of <50 dB and a speech discrimination score ≥50%, as determined by audiometric assessment. All patients had no sign of facial nerve dysfunction (House-Brackmann grade 1). Audiological and facial nerve examination were performed preoperatively as well as 1 week and 3 months postoperatively [18].

At the end of microsurgical resection, a 4-mm Flexible Video Endoscope (4 mm × 65 cm, Karl Storz, GmbH, Tuttlingen, Germany) was inserted into the surgical cavity, handled by the operator. The endoscope was introduced under both microscopic and endoscopic visualization to prevent injury to CPA structures, and the endoscopic tip was oriented into the IAC in order to detect tumor residue hiding in the deeper portion of IAC. If residual tumor was identified, microsurgical resection was pursued, and further endoscopic controls were repeated until complete tumor resection was accomplished [18].

Complete microsurgical resection was achieved in all cases. The endoscopic exploration of the IAC reportedly revealed, in all cases, a residual tumor in the lateral portion of IAC; therefore, multiple endoscopic controls and pursuit of further microsurgical resection of these endoscopically visualized residuals were attempted. Tumor residual fragments detected in the fundus of IAC were completely resected in all cases, as confirmed in the postoperative MRI. Hearing function was constantly monitored during surgery, and no variations of V-wave amplitude were registered. As expected, in the postoperative period, all patients maintained the preoperative hearing competence. Similarly, facial nerve was always anatomically and functionally preserved, and all patients showed a postoperative normal facial nerve function (House-Brackmann grade 1) [18, 19].

The main advantage of the flexible endoscope is the possibility to orient the endoscopic tip directly into the IAC to obtain an optimal visualization of the fundus; also, when introduced along the corridors between the cranial nerves, the device can be arranged in the configurational shape in order to obtain a safe corridor between the dorsal neurovascular structures of the posterior fossa [18].

The main disadvantage in using a flexible endoscope is the necessity to hold it with both hands [18]. In fact, the rigid endoscope can be held by the assistant in one hand as described for the "freehand endoscope holding technique" [20], routinely used in the surgical practice. On the contrary, the flexible endoscope has to be manipulated with two hands: one to insert the endoscope in the surgical field and the other to orient the endoscopic tip in the right position (Fig. 15.2). Nevertheless, this limitation can be easily managed by means of a good cooperation and synchronicity between the operator and the assistant.

In their cadaveric study, Baidya et al. [21] first demonstrated a flexible endoscope-assisted retrosigmoid approach to perform microsurgical resection of an artificial polymer tumor model resembling a medium-sized (15–20 mm of diameter) VS. The resection was performed by first creating a corridor by removing the lower portion of the tumor and then by inserting the flexible endoscope through the same corridor in order to accomplish early visualization and preservation of the acoustico-facial bundle. Early visualization of the facial and vestibular cochlear nerve complex led to unhindered removal of the tumor model [21]. This study demonstrated that the endoscopic-assisted microsurgical resection of a medium-sized VS was feasible in the authors' tumor model study emulating real surgery; therefore, the use of flexible endoscope should not be limited to the removal of ICVSs and may be experimented also on larger tumors.

Fig. 15.2 Bimanual holding of flexible endoscope

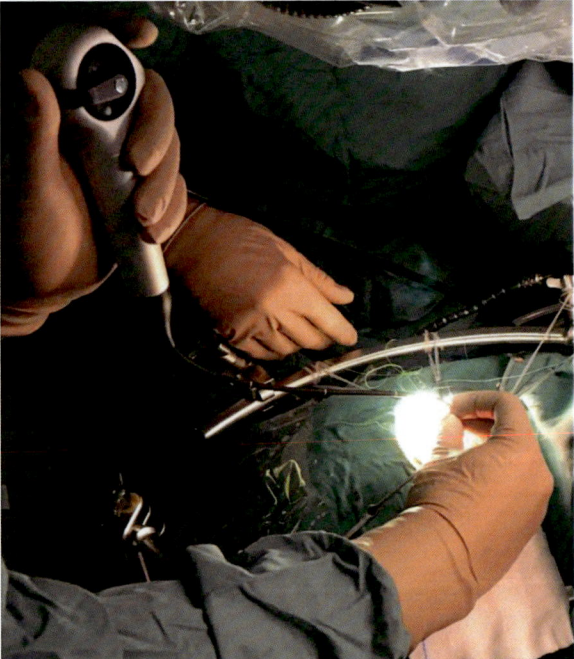

Further advances in endoscopic structure, such as the recently described ultrathin flexible endoscope with integrated irrigation and suction, could further improve the handling of surgical endoscopes and facilitate the visualization of the deep position of IAC [22].

References

1. Samii M, Matthies C. Management of 1000 vestibular schwannomas (acoustic neuromas): the facial nerve—preservation and restitution of function. Neurosurgery. 1997;40(4):684–94; discussion 94–5.
2. Nonaka Y, Fukushima T, Watanabe K, Friedman AH, Sampson JH, Mcelveen JT, et al. Contemporary surgical management of vestibular schwannomas: analysis of complications and lessons learned over the past decade. Neurosurgery. 2013;72(2 Suppl Operative):ons103–15; discussion ons15.
3. Myrseth E, Pedersen PH, Møller P, Lund-Johansen M. Treatment of vestibular schwannomas. Why, when and how? Acta Neurochir. 2007;149(7):647–60; discussion 60.
4. Pennings RJ, Morris DP, Clarke L, Allen S, Walling S, Bance ML. Natural history of hearing deterioration in intracanalicular vestibular schwannoma. Neurosurgery. 2011;68(1):68–77.
5. Wanibuchi M, Fukushima T, Zomordi AR, Nonaka Y, Friedman AH. Trigeminal schwannomas: skull base approaches and operative results in 105 patients. Neurosurgery. 2012;70(1 Suppl Operative):132–43; discussion 43–4.
6. Committee on Hearing and Equilibrium guidelines for the evaluation of hearing preservation in acoustic neuroma (vestibular schwannoma). American Academy of Otolaryngology-Head and Neck Surgery Foundation, INC. Otolaryngol Head Neck Surg. 1995;113(3):179–80.
7. Samii M, Metwali H, Gerganov V. Efficacy of microsurgical tumor removal for treatment of patients with intracanalicular vestibular schwannoma presenting with disabling vestibular symptoms. J Neurosurg. 2017;126(5):1514–9.
8. Mazzoni A, Zanoletti E, Denaro L, Martini A, Avella D. Retrolabyrinthine meatotomy as part of retrosigmoid approach to expose the whole internal auditory canal: rationale, technique, and outcome in hearing preservation surgery for vestibular schwannoma. Oper Neurosurg (Hagerstown). 2018;14(1):36–44.
9. Pillai P, Sammet S, Ammirati M. Image-guided, endoscopic-assisted drilling and exposure of the whole length of the internal auditory canal and its fundus with preservation of the integrity of the labyrinth using a retrosigmoid approach: a laboratory investigation. Neurosurgery. 2009;65(6 Suppl):53–9; discussion 9.
10. Fukushima T. Endoscopy of Meckel's cave, cisterna magna, and cerebellopontine angle. Technical note. J Neurosurg. 1978;48(2):302–6.
11. Kurucz P, Baksa G, Patonay L, Thaher F, Buchfelder M, Ganslandt O. Endoscopic approachroutes in the posterior fossa cisterns through the retrosigmoid keyhole craniotomy: an anatomical study. Neurosurg Rev. 2017;40(3):427–48.
12. Takemura Y, Inoue T, Morishita T, Rhoton AL. Comparison of microscopic and endoscopic approaches to the cerebellopontine angle. World Neurosurg. 2014;82(3–4):427–41.
13. Abolfotoh M, Bi WL, Hong CK, Almefty KK, Boskovitz A, Dunn IF, et al. The combined microscopic-endoscopic technique for radical resection of cerebellopontine angle tumors. J Neurosurg. 2015;123(5):1301–11.
14. Chovanec M, Zvěřina E, Profant O, Skřivan J, Cakrt O, Lisý J, et al. Impact of video-endoscopy on the results of retrosigmoid-transmeatal microsurgery of vestibular schwannoma: prospective study. Eur Arch Otorhinolaryngol. 2013;270(4):1277–84.
15. Göksu N, Bayazit Y, Kemaloğlu Y. Endoscopy of the posterior fossa and dissection of acoustic neuroma. J Neurosurg. 1999;91(5):776–80.

16. Tatagiba MS, Roser F, Hirt B, Ebner FH. The retrosigmoid endoscopic approach for cerebellopontine-angle tumors and microvascular decompression. World Neurosurg. 2014;82(6 Suppl):S171–6.
17. Turek G, Cotúa C, Zamora RE, Tatagiba M. Endoscopic assistance in retrosigmoid transmeatal approach to intracanalicular vestibular schwannomas—an alternative for middle fossa approach. Technical note. Neurol Neurochir Pol. 2017;51(2):111–5.
18. Corrivetti F, Cacciotti G, Scavo CG, Roperto R, Mastronardi L. Flexible endoscopic-assisted microsurgical radical resection of intracanalicular vestibular schwannomas by retrosigmoid approach: operative technique. World Neurosurg. 2018;115:229–33.
19. House JW, Brackmann DE. Facial nerve grading system. Otolaryngol Head Neck Surg. 1985;93(2):146–7.
20. de Divitiis O, Cavallo LM, Dal Fabbro M, Elefante A, Cappabianca P. Freehand dynamic endoscopic resection of an epidermoid tumor of the cerebellopontine angle: technical case report. Neurosurgery. 2007;61(5 Suppl 2):E239–40; discussion E40.
21. Baidya NB, Berhouma M, Ammirati M. Endoscope-assisted retrosigmoid resection of a medium size vestibular schwannoma tumor model: a cadaveric study. Clin Neurol Neurosurg. 2014;119:35–8.
22. Otani N, Morimoto Y, Fujii K, Toyooka T, Wada K, Mori K. Flexible ultrathin endoscope integrated with irrigation suction apparatus for assisting microneurosurgery. World Neurosurg. 2017;108:589–94.

Fluid Cement for Bone Closure

16

Alberto Campione, Guglielmo Cacciotti,
Raffaelino Roperto, Carlo Giacobbo Scavo,
and Luciano Mastronardi

Hydroxyapatite is the primary constituent of human bone and is composed of a calcium phosphate mineral compound $[Ca(PO_4)_2(OH)_2]$. Porous ceramic hydroxyapatite preparations were historically derived from calcium carbonate skeletons of sea coral; ceramic preparations would be produced by fusing individual crystals by heat, thus resulting in an already hard, brittle, and nonresorbable material [1, 2]. Today, nonceramic HAC is produced synthetically by crystallization of hydroxyapatite at a physiological pH through an isothermic reaction. Intraoperative combination of tetracalcium phosphate and dicalcium phosphate anhydrous with an aqueous solution of sodium phosphate forms a material that can be easily sculpted and hardens within 5–10 min. After 4 h, this formulation coverts to hydroxyapatite cement (HAC) and is no longer water soluble. HAC is available in a thick granular paste (BoneSource, Stryker, Kalamazoo, MI, USA) and a thickened injectable form (HydroSet, Stryker, Kalamazoo, MI, USA). The properties of the two forms are similar, with HydroSet having features that are slightly more resistant to the effects of moisture. The choice is based on the type of defect and surgeon preference; BoneSource is a thicker paste and easier to handle when sculpting a large defect. HydroSet is a thick liquid formulation and may be easier to use to fill narrow crevices. It remains invaluable for the treatment of discontinuous full-thickness crater defects where it can be used to preferentially fill in depressions [1, 3].

HAC is not osteogenic, but it is osteoconductive in that it serves as a scaffold over and inside which bone can grow without intervening fibrous tissue. Several studies demonstrated long-term replacement of implanted HAC with new cortical and trabecular bone growth through biopsies and serial radiographic scans [2, 4–6]. Unlike many other alloplasts, when HAC sets properly, it does not result in sustained inflammation, toxicity, foreign body giant cell reactions, fibrous capsule

A. Campione (✉) · G. Cacciotti · R. Roperto · C. Giacobbo Scavo · L. Mastronardi
Department of Neurosurgery, San Filippo Neri Hospital—ASLRoma1, Rome, Italy
e-mail: mastro@tin.it

© Springer Nature Switzerland AG 2019
L. Mastronardi et al. (eds.), *Advances in Vestibular Schwannoma Microneurosurgery*, https://doi.org/10.1007/978-3-030-03167-1_16

formation, or abnormalities in calcium phosphate metabolism [1, 2, 7]. HAC in both forms must be applied in a dry field either directly on bone in a partial-thickness defect or filling a full-thickness defect abutting the native bone to permit osteocon-duction and osteointegration [3].

With micropore diameters of 2–5 nm, HAC is highly resistant to infection. The majority of infections associated with the use of HAC are related to secondary infection after seroma. In the early postoperative period, moisture leads to failure of the particles to set, migration of the particles, and eventual seroma formation [3]. Such cases were first reported by Kveton et al. [2]: two patients undergoing recon-struction of suboccipital craniectomy defects received radiographical examination that revealed total cement resorption. The authors addressed a bloody field during closure and subsequent hematoma as the cause of HAC resorption [1, 2], thus con-firming it is a moisture-sensible material. Similarly, in the delayed setting, trauma causing fracture of the HAC can result in prolonged edema and delayed seroma with consequent disintegration of the material or superinfection [3].

Fortunately, with proper application and patient counseling, complications with the use of HAC remain less than 5% in the hands of experienced surgeons. Proper application, including the dryness of the field, is critical to achieving a successful outcome, especially in dependent sites at the skull base which are prone to hema-toma formation and exposure to constant dural pulsation [2]. HAC must be applied to the surgical site immediately after thoroughly mixing with sodium phosphate solution. Manipulation should be performed within the first few minutes, allowing ample time for the material to set undisturbed and uncovered [3].

Whether HAC should be used as the sole bone substitute during cranioplasty or as an adjunct to other materials, it is still a matter of debate, and no direct answer really exists. The employment of HAC depends on the type of defect that is to be restored, the conditions wherein the cranioplasty is performed, and obviously the dimension of the defect itself. Tadros and Costantino [3] provided a detailed algo-rithm to guide cranial reconstruction of acquired defects. Based on the authors' indications, when the defect only regards bone and the soft tissues are preserved, a titanium mesh should be used and associated with duraplasty in case of prior tumor resection. As an alternative, HAC was proposed in the case of a single <5 cm^2 defect, while the association of both the mesh and HAC would be best reserved to larger bony defects [3]. In the case of vestibular schwannoma (VS) surgery performed via retrosigmoid approach, the area of the craniotomy is approximately 3 cm^2 and the sole HAC would be recommended; indeed, many studies reported the outcomes of retrosigmoid cranioplasty with only HAC.

Reports of HAC in retrosigmoid craniotomy are scarce and inconsistent, with some groups reporting success and others reporting unacceptably high complication rates [8–10]. Even fewer are the reports of HAC in retrosigmoid craniotomy after VS resection: Kveton et al. [2] reported successful reconstruction of the cranial defect in five of seven patients. Total resorption of HAC—probably due to wound closure without proper hemostasis—occurred in the other two patients. The postop-erative CSF leak rate was not reported; as far as wound infections and meningitis were concerned, one patient developed aseptic meningitis. Interestingly, when compared to patients who had not received reconstructive cranioplasty, the ones enrolled in the study referred a lower incidence of postoperative headache, −20% vs. 60%, respectively.

More recent studies focused on the outcomes of retrosigmoid cranioplasty in the setting of microvascular decompressions (MVD) or cranial nerve disturbances—only a minority of which involved VSs. Eseonu et al. [9] evaluated the outcomes of postoperative CSF leak and wound infection for patients undergoing a complete cranioplasty using calcium phosphate cement versus incomplete cranioplasty using polyethylene titanium mesh following a retrosigmoid craniectomy for MVD. According to the authors' definitions, "complete" cranioplasty is the replacement of the entire calvarial defect, using either the bone flap and a bone analog together or a bone analog alone; an "incomplete" cranioplasty involves partial reconstruction of the calvarial defect, using only a bone flap, a titanium mesh, or nothing at all. One hundred five patients underwent a complete cranioplasty (R group), and 116 patients received a polyethylene titanium mesh for incomplete cranioplasty (NR group); CSF leak rate was found significantly different ($p = 0.03$) between the groups, as low as 0% in the R group and as high as 4.5% in the NR group. No statistically significant differences were reported in postoperative wound infection rate, 2% vs. 2.7% in the R group and in the NR group, respectively.

Aldahak et al. [11] aimed to determine the safety and effectiveness of HAC for reconstruction of retrosigmoid craniotomy used for treatment of various cranial nerves disorders. Ninety-three patients were included in analysis, and HAC cranioplasty without additional materials was accomplished in all cases. While there were no deep-seated postoperative infections, three patients (3.2%) developed superficial wound complications requiring revision. CSF external leak from the wound did not occur in any cases; however, one patient (1%) presented a pseudomeningocele 15 days postoperatively and underwent a surgical revision after an external lumbar drainage failure; the HAC was changed to a titanium mesh cranioplasty. Long-term incisional pain occurred in one case (1%) where prescription pain medication was continually requested by the patient. The pain subsided to a low-intensity tenderness at the operative site at 6 months but remained present at 1 year. One patient (1%) was dissatisfied with the cosmetic results due to the scar that formed after revision for stitch abscess. There were no complaints of palpable or visible deformities beneath the skin. All the other patients stated complete satisfaction with the cosmetic results [11].

Luryi et al. [10] report rates of cerebrospinal fluid leak, wound infection, and other complications after repair of retrosigmoid craniotomy with HAC. The indications for the surgical procedures were diverse; 20 cases were identified, 5 of which had VSs. No instances of CSF leak were eventually noted; no infections occurred in the immediate postoperative period. However, one patient (5%) developed a cyst near the postauricular incision 4 months postoperatively; this cyst progressed over the following several months and subsequently erupted, leading to infection. Imaging demonstrated a contiguous cerebellar lesion which required operative removal and repair.

Among the five patients with VSs, 1 (20%) suffered from middle ear effusion, and another suffered a fall 7 days postoperatively and was found to have a small cerebellar hemorrhage; this patient was monitored and improved without operative intervention [10].

Benson and Djalilian [8] describe two cases of hydroxyapatite resorption and subsequent seroma formation in patients who had undergone retrosigmoid craniotomy. The presentation in both cases mimicked a CSF leak. In both cases, the fragmented cement was removed, and the patients experienced no further complications. The authors concluded that HAC should not be used for the reconstruction of retrosigmoid/suboccipital craniotomies because it was associated with unacceptably high complication rates [8].

At our institution, bone flap is always repositioned and fixed with titanium screws prior to layering the HAC (Fig. 16.1). HydroSet (Stryker, Kalamazoo, MI, USA) and OsteoVation (Osteomed, Addison, TX, USA) have been used as HAC cements; both are injectable materials that have to be mixed at the moment and share the fundamental biomechanical properties. OsteoVation, though, is chemically formulated to set in a wet field environment eliminating the need to meticulously dry the

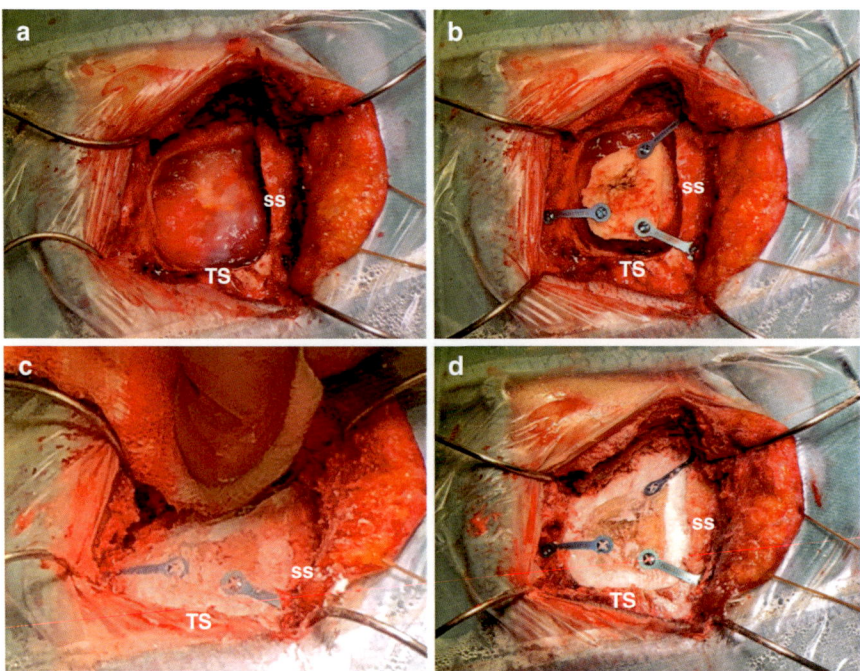

Fig. 16.1 Application of hydroxyapatite bone cement (HAC). In every picture, the sigmoid sinus and the transverse sinus are indicated. *SS* sigmoid sinus, *TS* transverse sinus. (**a**) The surgical field of a left retrosigmoid craniotomy is shown after vestibular schwannoma resection, dural closure with duraplasty, and fibrin sealant application. (**b**) The bone flap has been repositioned and fixed with titanium screws. (**c**) The HAC is being applied and modeled to achieve the best cosmetic outcome. (**d**) The HAC has been successfully layered resulting in optimal bone void filling

operative site prior to implantation; in addition, its chemical composition is slightly different than that of HydroSet, as it is a compound of alpha-tricalcium phosphate and sodium silicate. As reconstructive retrosigmoid cranioplasty was combined with an underlay hourglass-shaped autologous pericranium duraplasty in a recent study of ours, no postoperative wound infections or meningitis were observed nor were clinically significant cerebrospinal fluid (CSF) leaks reported [12].

Further studies with multi-institutional participation are needed to better elucidate the advantages and complication rates of HAC retrosigmoid cranioplasty in VS surgery [10]. However, in the perspective of reducing the postoperative CSF leak rate as much as possible, it is advisable to focus the upcoming research not only on cranioplasty but also on the combination duraplasty-cranioplasty, which yielded very good results in our experience.

References

1. Kveton JF, Coelho DH. Hydroxyapatite cement in temporal bone surgery: a 10 year experience. Laryngoscope. 2004;114(1):33–7.
2. Kveton JF, Friedman CD, Piepmeier JM, Costantino PD. Reconstruction of suboccipital craniectomy defects with hydroxyapatite cement: a preliminary report. Laryngoscope. 1995;105(2):156–9.
3. Tadros M, Costantino PD. Advances in cranioplasty: a simplified algorithm to guide cranial reconstruction of acquired defects. Facial Plast Surg. 2008;24(1):135–45.
4. Costantino PD, Friedman CD, Jones K, Chow LC, Pelzer HJ, Sisson GA. Hydroxyapatite cement. I. Basic chemistry and histologic properties. Arch Otolaryngol Head Neck Surg. 1991;117(4):379–84.
5. Friedman CD, Costantino PD, Jones K, Chow LC, Pelzer HJ, Sisson GA. Hydroxyapatite cement. II. Obliteration and reconstruction of the cat frontal sinus. Arch Otolaryngol Head Neck Surg. 1991;117(4):385–9.
6. Kveton JF, Friedman CD, Costantino PD. Indications for hydroxyapatite cement reconstruction in lateral skull base surgery. Am J Otol. 1995;16(4):465–9.
7. Kamerer DB, Hirsch BE, Snyderman CH, Costantino P, Friedman CD. Hydroxyapatite cement: a new method for achieving watertight closure in transtemporal surgery. Am J Otol. 1994;15(1):47–9.
8. Benson AG, Djalilian HR. Complications of hydroxyapatite bone cement reconstruction of retrosigmoid craniotomy: two cases. Ear Nose Throat J. 2009;88(11):E1–4.
9. Eseonu CI, Goodwin CR, Zhou X, Theodros D, Bender MT, Mathios D, et al. Reduced CSF leak in complete calvarial reconstructions of microvascular decompression craniectomies using calcium phosphate cement. J Neurosurg. 2015;123(6):1476–9.
10. Luryi AL, Bulsara KR, Michaelides EM. Hydroxyapatite bone cement for suboccipital retrosigmoid cranioplasty: a single institution case series. Am J Otolaryngol. 2017;38(4):390–3.
11. Aldahak N, Dupre D, Ragaee M, Froelich S, Wilberger J, Aziz KM. Hydroxyapatite bone cement application for the reconstruction of retrosigmoid craniectomy in the treatment of cranial nerves disorders. Surg Neurol Int. 2017;8:115.
12. Mastronardi L, Cacciotti G, Caputi F, Roperto R, Tonelli MP, Carpineta E, et al. Underlay hourglass-shaped autologous pericranium duraplasty in "key-hole" retrosigmoid approach surgery: technical report. Surg Neurol Int. 2016;7:25.

Aspirin Administration for Control of Tumor Millimetric Residual

Alberto Campione, Guglielmo Cacciotti, Raffaelino Roperto, Carlo Giacobbo Scavo, and Luciano Mastronardi

The emerging therapies in vestibular schwannoma (VS) surgery are the result of a large amount of translational research studies conducted over the last years. Three main fields have been explored: medical therapy, prehabilitation, and frontiers in surgical care, as outlined by the latest guidelines edited by the Congress of Neurological Surgeons (CNS) [1]. As regards medical therapy, new details about the role of proinflammatory pathways in the molecular pathogenesis of VSs have inspired the researchers to experiment aspirin and nonsteroidal anti-inflammatory drugs (NSAIDs) in the treatment of the tumor. The evidence-based guidelines on emerging therapies by the CNS recommend that aspirin administration may be considered for use in patients undergoing observation to prevent tumor growth [1]; however, controversial results have been reported in the recent literature. The effects of salicylates and NSAIDs on VS growth rate have been studied both in vitro and in vivo.

17.1 In Vitro Studies

In 2011, Hong et al. [2] examined 30 patients with VS, 15 of which were sporadic, for cyclooxygenase 2 (COX-2) expression, microvessel density, and proliferation rate by immunohistochemical methods. The COX enzymes catalyze the biosynthesis of prostaglandins (PGs), hormonelike lipid compounds that can trigger the inflammatory response. The authors found that COX-2 expression was detected in all but one patient. In addition, they reported that the Ki-67 proliferation index in VSs with high COX-2 expression was significantly larger when compared with VSs with weak COX-2 expression; therefore, they concluded that COX-2 may play an integral role in VS proliferation and that expression of COX-2 correlated with VS growth rate [2].

A. Campione (✉) · G. Cacciotti · R. Roperto · C. Giacobbo Scavo · L. Mastronardi
Department of Neurosurgery, San Filippo Neri Hospital—ASLRoma1, Rome, Italy
e-mail: mastro@tin.it

© Springer Nature Switzerland AG 2019
L. Mastronardi et al. (eds.), *Advances in Vestibular Schwannoma Microneurosurgery*, https://doi.org/10.1007/978-3-030-03167-1_17

In 2015, Dilwali et al. [3] investigated the role of COX-2 in VSs and tested COX-2-inhibiting salicylates against VSs. In the study, the efficacy of three different salicylates was assessed, namely, aspirin, sodium salicylate (NaSal), and 5-aminosalicylic acid (5-ASA).

COX-2 was found to be aberrantly expressed in human VS specimens and primary human VS cells in comparison with control human nerve specimens and primary Schwann cells (SCs), respectively. Furthermore, levels of prostaglandin E_2 (PGE_2), the downstream enzymatic product of COX-2, were correlated with primary VS culture proliferation rate. The significant correlation of PGE_2 levels with VS culture proliferation rate further confirmed the previous finding that COX-2 expression correlated with VS growth rate [2]. On the contrary, substantially decreased PGE_2 levels in the media after salicylate treatment suggested that the salicylates led to COX-2 inhibition. In addition, aspirin and NaSal can also inhibit nuclear factor kappa-light-chain-enhancer of activated B cells (NF-kB) directly, through blockade of I kappa B kinase (IkK); interestingly, the COX-2 gene promoter has a kB binding site. Thus, Dilwali et al. speculated that the inhibition of NF-kB-driven cell proliferation due to salicylates might eventually result in a decrease in COX-2 expression [3].

The tested drugs neither increased VS cell death nor affected healthy SCs. The cytostatic effect of aspirin in vitro was in line with a previous clinical finding by Kandathil et al. [4] that patients with VS taking aspirin demonstrated reduced tumor growth [3]. The cytostatic effect of salicylates against VS cells seemed to be specific to the neoplastic cells because treating healthy SCs with the same concentrations of the drugs did not lead to a decrease in cell proliferation.

17.2   Clinical Studies

Encouraging evidences were reported in the clinical studies by Kandathil et al. [4, 5], which served as the core basis for the elaboration of the guidelines edited by the CNS. However, more recent studies [6, 7] have reported controversial results that underline how important and useful a double-blind, placebo-controlled, randomized trial would be to determine efficacy with more certainty [4–7].

In 2014, Kandathil et al. [4] explored the role of aspirin in minimizing sporadic VS growth (defined as >0 mm/year change in serial MRI scans conducted at least 4 months apart) in vivo. Three hundred forty-seven patients were followed by serial MRI scans. Eighty-one patients took aspirin, 33 (40.7%) of which demonstrated tumor growth, while 48 (59.3%) did not. Of the 266 non-aspirin users, 154 (57.9%) demonstrated VS growth and 112 (42.1%) did not. The difference in VS growth versus no growth between aspirin users and non-users was statistically significant ($p = 0.0076$; odds ratio (OR), 0.50; 95% confidence interval (CI), 0.29–0.85) and was not confounded by age or sex. The authors demonstrated, for the first time, inverse association between aspirin intake and VS growth rate, categorized as growth versus no growth [4].

In 2016, the same group [5] evaluated this association using more accurate tumor volumetric measurements to detect tumor growth (defined as >20% positive change in volume from the first MRI scan). Eighty-six patients underwent sequential MRI scans suitable for 3D-segmented volumetric analysis for up to 11 years of follow-up (median 53 months). Twenty-five had documented history of aspirin intake; 8 (32%) of them demonstrated VS growth, while 17 (68%) patients did not. Of the 61 non-aspirin users, 36 (59%) demonstrated growth. A significant inverse association was found between aspirin intake and VS growth: $p = 0.03$; OR, 0.32; and 95% CI, 0.11–0.91; again, the result was not confounded by age or sex. On the basis of such results, the authors suggested that aspirin may represent a noninvasive cytostatic pharmacological approach to the treatment of VSs. They speculated that such approach could be considered to prevent the need for surgery or radiation therapy in the best case; as an alternative, aspirin may at least allow the patient and clinician more time to plan the intervention [5].

Hunter et al. [6] first reported data clearly in contrast to that previously observed by Kandathil et al. [4, 5]. Of 564 patients with VS who underwent at least two MRI scans before intervention, 158 were aspirin users, 96 were NSAID users, and 20 took both the types of medications, at different dosages. Neither aspirin use nor aspirin dosage was associated with VS tumor growth (defined as ≥ 2 mm increase in the maximum tumor diameter between consecutive MRI studies), presenting tumor diameter or mean VS growth rate. Further, neither non-aspirin NSAIDs intake nor the degree of COX-2 selectivity showed any significant correlation with the primary outcomes of the study [6]. The authors addressed the different study designs, follow-up periods, and tumor growth definitions as the likely cause of such discrepancies in relation to the studies by Kandathil et al. [4, 5].

MacKeith et al. [7] conducted a study which utilized a postal questionnaire and telephone interviews to determine aspirin exposure. Propensity score matching was used to control for age, sex, and tumor size. Cases were defined as patients with VS proven to have grown on serial MRI scans; on the contrary, controls were defined as patients with stable VS. Two hundred twenty cases and 217 controls were eventually enrolled in the study; aspirin exposure was more common in stable than in growing VSs (22.1% vs. 17.3%). However, following matching to control for covariates, aspirin was not found to be associated with VS stability ($p = 0.475$). Multiple logistic regression (analysis of variance) found tumor size at diagnosis to be the only factor strongly associated with tumor growth (defined as in Hunter et al. [6]) ($p < 0.0001$). The authors compared their study to those by Kandathil et al. [4, 5] and found many differences in the definition of tumor growth and aspirin exposure that could have led to different results. Foremost, MacKeith et al. reported that similar results to those of the Kandathil studies had been obtained (with an inverse association between aspirin intake and VS growth) before tumor size stratification of the case and control groups. However, following propensity score matching to control for the difference in tumor size, no association between aspirin use and tumor stability was demonstrated. This latter could be the most likely explanation for the conflicting results [7].

17.3 Future Perspectives and Personal Experience

A double-blind, placebo-controlled, randomized trial should be warranted to further characterize the role of aspirin in the treatment of VSs. Aspirin has indeed many advantages in that it is a commonly used drug whose pharmacokinetics and pharmacodynamics have been widely studied—i.e., efficacy, collateral effects, and toxicity are well known and predictable.

The first question to be solved would obviously be that of efficacy as a cytostatic on VSs; should it be confirmed, further research would be needed to precisely determine the correct posology. In fact, aspirin and, more generally, NSAID concentrations effective against VSs in vivo have not been established yet. Dilwali et al. [3] speculated that, on the basis of the concentration of salicylates found in the culture media in their study, doses of 800 mg of aspiring should be effective in vivo. However, this kind of speculation oversimplifies the complex process of drug distribution in the cerebrospinal fluid (CSF). As a matter of fact, salicylates readily cross the blood-brain barrier and can reach up to 50% of the concentration present in the blood [8], an appealing aspect that makes translation of salicylates against VS even more promising. Regardless, salicylate concentrations in tumor tissue are likely to be similar to those in serum because the blood-brain barrier is compromised in intracranial tumors [9].

Second and even more ambitious target would be application of aspirin as a cytostatic for secondary prevention of tumor growth. In other words, using aspirin as a cytostatic in the context of a postoperative millimetric tumor residue would hopefully lower the risk of recurrences. In the personal experience of one of the senior authors of this book (L.M.), nine patients have undergone subtotal VS resection and have been prescribed postoperative aspirin off-label with the purpose of avoiding recurrences. As the index patient of the series was operated in June 2014, the maximum follow-up period has been 47 months, and only the first patient was eventually affected by a tumor regrowth needing reoperation; the remaining eight patients are being followed up and show stable tumors [data not published].

References

1. Van Gompel JJ, Agazzi S, Carlson ML, Adewumi DA, Hadjipanayis CG, Uhm JH, et al. Congress of Neurological Surgeons systematic review and evidence-based guidelines on emerging therapies for the treatment of patients with vestibular schwannomas. Neurosurgery. 2018;82(2):E52–E4.
2. Hong B, Krusche CA, Schwabe K, Friedrich S, Klein R, Krauss JK, et al. Cyclooxygenase-2 supports tumor proliferation in vestibular schwannomas. Neurosurgery. 2011;68(4):1112–7.
3. Dilwali S, Kao SY, Fujita T, Landegger LD, Stankovic KM. Nonsteroidal anti-inflammatory medications are cytostatic against human vestibular schwannomas. Transl Res. 2015;166(1):1–11.
4. Kandathil CK, Dilwali S, Wu CC, Ibrahimov M, McKenna MJ, Lee H, et al. Aspirin intake correlates with halted growth of sporadic vestibular schwannoma in vivo. Otol Neurotol. 2014;35(2):353–7.

5. Kandathil CK, Cunnane ME, McKenna MJ, Curtin HD, Stankovic KM. Correlation between aspirin intake and reduced growth of human vestibular schwannoma: volumetric analysis. Otol Neurotol. 2016;37(9):1428–34.
6. Hunter JB, O'Connell BP, Wanna GB, Bennett ML, Rivas A, Thompson RC, et al. Vestibular schwannoma growth with aspirin and other nonsteroidal anti-inflammatory drugs. Otol Neurotol. 2017;38(8):1158–64.
7. MacKeith S, Wasson J, Baker C, Guilfoyle M, John D, Donnelly N, et al. Aspirin does not prevent growth of vestibular schwannomas: a case-control study. Laryngoscope. 2018;128(9):2139–44.
8. Bannwarth B, Netter P, Pourel J, Royer RJ, Gaucher A. Clinical pharmacokinetics of non-steroidal anti-inflammatory drugs in the cerebrospinal fluid. Biomed Pharmacother. 1989;43(2):121–6.
9. Bart J, Groen HJ, Hendrikse NH, van der Graaf WT, Vaalburg W, de Vries EG. The blood-brain barrier and oncology: new insights into function and modulation. Cancer Treat Rev. 2000;26(6):449–62.

DTI for Facial Nerve Preoperative Prediction of Position and Course

18

Alberto Campione, Guglielmo Cacciotti,
Raffaelino Roperto, Carlo Giacobbo Scavo,
and Luciano Mastronardi

The goal of modern-day vestibular schwannoma (VS) surgery is total tumor removal with the preservation of neurological function and quality of life. Postoperative facial nerve (N VII) paralysis is one of the major complications of VS surgery and is due to anatomical interruption of N VII during surgical procedure in 5% of VS patients. Therefore, attempts to preoperatively determine the course of N VII assume great significance. This is especially true for larger VSs (tumor size >3 cm), wherein N VII preservation becomes increasingly difficult because of the unpredictable displacement of the nerve within the cerebellopontine angle (CPA); in such cases, the nerve is often flattened or splayed, making its identification particularly demanding for the surgeon [1]. The current strategy in VS surgery dictates early intraoperative identification of N VII based on its relation to anatomical landmarks and especially by means of electrical stimulation and electromyographic monitoring [1]. Any imaging study preoperatively demonstrating the course of N VII in VS patients should theoretically enhance surgical safety, enabling the surgeon to avoid unexpected injuries to the nerve [2].

Sartoretti-Schefer et al. demonstrated that it is very difficult to visualize the course of N VII using conventional MRI techniques such as MR cisternography in case of large tumors (>2.5 cm) because of the focal nerve thinning and the obliteration of landmarks occurring within the internal auditory canal (IAC) and CPA [3]. The use of diffusion tensor imaging-fiber tracking (DTI-FT, also known as diffusion tensor tractography, DTT) for N VII tracking has evolved as a reliable technique in this regard.

DTT is a novel modality of MRI analysis that measures the diffusion direction of water molecules by combining multiple diffusion-weighted scans taken from multiple gradient directions [4]. The diffusion of water molecules is thought to be anisotropic inside white matter tracts and therefore maximal along the direction of the fiber tracts. DTT allows the 3D reconstruction of the cranial nerves in healthy individuals. A 3D

A. Campione (✉) · G. Cacciotti · R. Roperto · C. Giacobbo Scavo · L. Mastronardi
Department of Neurosurgery, San Filippo Neri Hospital—ASLRoma1, Rome, Italy
e-mail: mastro@tin.it

© Springer Nature Switzerland AG 2019
L. Mastronardi et al. (eds.), *Advances in Vestibular Schwannoma
Microneurosurgery*, https://doi.org/10.1007/978-3-030-03167-1_18

vector field (tensor) is assigned to each voxel. This information is then used to reconstruct and represent pictorially the white matter tracts within a specific region of interest (ROI). The reconstructed model of the fiber tracts is obtained in a highly reproducible manner [5]. DTT reconstruction of N VII fibers in case of VS is considered successful if a continuous tract of fibers is seen to extend from the internal auditory meatus to the brainstem along the tumor capsule of a VS [6].

18.1 Diffusion Tensor Imaging-Based Fiber Tracking: State of the Art

In 2006, Taoka et al. [2] first used DTT to preoperatively visualize the course of N VII displaced by VS and evaluated the concordance between radiological predictions and surgical findings. Eight patients with VS were enrolled who had undergone surgery. The authors obtained a tract that connected the internal auditory meatus and the brainstem, which was considered to represent the facial nerve in 7 (87.5%) of eight cases. The course of the constructed tract agreed with surgical findings in 5 (71.4%) of those 7 cases [2].

Gerganov et al. [1] described a series of 22 consecutive patients with large VSs who had DTI scans acquired and postprocessed with navigational software to obtain N VII fiber tracking. Fibers corresponding to the anatomical location and course of N VII from the brainstem to the internal auditory meatus were identified with the DTT technique in all 22 (100%) cases. The surgical concordance rate (SCR)—i.e., the rate of agreement between imaging predictions and surgical findings—was 90.9%. No correlation was found between DTT results and the two morphological types of the nerve (compact or flat). In addition to a pictorial display of the tracts, DTT provides information on specific parameters, including fractional anisotropy (FA), which is the degree of water diffusion restrictedness of structures such as axons [7]. A FA analysis can relay information regarding the microstructure and axonal properties of a nerve, including demyelination, inflammation, and axon diameter. Moreover, quantitative measurements of FA can be used to study axonal integrity values of fiber tracts. On this basis, Zhang et al. [8] postulated that maximal FA of N VII reflects the nerve properties to a certain extent, including the morphological characteristics. In their 30-patient prospective series, the maximal FA of N VII demonstrated moderate diagnostic performance in distinguishing compact from flat N VIIs (AUC = 0.84; 95% confidence interval (CI), 0.69–0.98; $p = 0.002$), as opposed to previous findings by Gerganov et al. In addition, Zhang et al. reported successful DTT-based preoperative visualization of N VII in all 30 (100%) patients; the SCR was as high as 96.7% [8].

Choi et al. [6] prospectively collected data from 11 patients with VS, who underwent preoperative DTT for N VII. Imaging results were correlated with intraoperative findings. Facial nerve courses on preoperative tractography were entirely correlated with intraoperative findings in all patients (SCR = 100%). The authors also first used postoperative DTT 3 months after surgery to confirm N VII anatomical preservation, which was reported in all patients [6].

Wei et al. [9] enrolled 23 consecutive patients with VS of stage T3b to T4b according to the Hannover classification. The DTT technique was used to preoperatively identify—and thus predict—the position of both N VII and cochlear nerve. As regards N VII, its visualization on DTT scans was feasible for all the patients (100%), and a SCR as high as 91.3% was reported. The identification of the cochlear nerve was more cumbersome: fibers of unclear function were regarded as belonging to it based on relationships with regional landmarks in four patients with functional hearing [9].

N VII detection rate on DTT scans and SCR values reported by more recent studies are in line with the previous literature: Song et al. [10] described a series of 15 patients and reported preoperative visualization of N VII in 93.3% of the patients, with a SCR of 92.9%. Hilly et al. [11] performed a study wherein the DTI technique was established in 113 patients without tumors and in 21 patients with medium- and large-sized CPA tumors, treated surgically via a translabyrinthine approach. N VII was successfully preoperatively detected in 95% and 97% of patients with and without tumors, respectively; the SCR in operated patients was 90% [11].

Diverse systematic reviews have been published in the attempt to summarize the now considerable amount of studies concerning the supposed advantages of DTT technique [5, 12]. The latest one was performed by Savardekar et al. [5], who selected 14 studies on preoperative N VII localization in relation to a VS using the DTI-FT technique; the included articles were required to report the SCR, based on tracking results confirmation by using microscopic observation and electrophysiological monitoring during microsurgery or neuronavigation. All of the previously cited studies were included in this review. The authors calculated that, of the 234 VS patients who constituted the population of their pooled analysis, complete tracking of the nerve's course was obtained in 226 patients (96.6%). Surgical concordance with the preoperative DTI-FT findings was obtained in 205 patients (90.7%). The authors deemed preoperative DTT for N VII identification a useful adjunct in the surgical planning for large VSs (>2.5 cm) [5].

The guidelines on the role of imaging in the diagnosis and management of patients with VSs edited by the Congress of Neurological Surgeons [13] dictate that advanced T2-weighted MRI scan sequences (e.g., CISS/FIESTA or DTI) may be used to augment visualization of N VII course as part of preoperative evaluation. However, larger studies are required to look at the direct benefit offered by DTT in preserving postoperative facial function.

18.2 From the Vestibulo-Facial Complex to Facial Nerve Alone

Based on the definition of a successful DTT reconstruction of N VII fibers in case of VS as a continuous tract of fibers extending from the internal auditory meatus to the brainstem along the tumor capsule of a VS [6], it can be assumed that such tract would correspond to the vestibulo-facial complex, rather than to N VII alone. The application of tractography to detailed analysis of the cranial nerves is a recent technology; therefore, it has potential limitations. These include a limited ability to distinguish the fibers of VII and VIII nerves. Some information can be gained from the location of the fibers; however, individual fibers cannot be properly distinguished. The main technical limitations in this area are related to the size and close anatomic proximity of the facial and vestibular nerves, as well as limitations on voxel size. The consequence of these limitations is that these cranial nerves cannot be imaged separately on DTT scans [4].

Roundy et al. [14] first developed and used a new high-density diffusion tensor imaging (HD-DT imaging) method, aimed at preoperatively identifying the location and course of the facial nerve in relation to large CPA (>2.5 cm) tumors. The authors prospectively studied five patients who underwent preoperative traditional standard- and HD-DT imaging. Imaging results were correlated with intraoperative findings. Utilizing their HD-DT imaging method, the authors positively identified the location and course of the facial nerve in all patients (SCR = 100%). In contrast, using a standard DT imaging method, the authors were unable to identify the facial nerve in four of the five patients [14].

A similar high-definition technique was used by Yoshino et al. [15] who proposed that a substantial improvement in image resolution could be achieved with high-angle diffusion magnetic resonance imaging and atlas-based fiber tracking to provide detailed trajectories of the cranial nerves. Five neurologically healthy adults and three patients with brain tumors were scanned with diffusion spectrum imaging that allowed high-angular-resolution fiber tracking. In addition, a 488-subject diffusion MRI template constructed from the Human Connectome Project data was used to conduct atlas space fiber tracking of the nerves. The intrabrainstem portion of N VII could be traced from the root exit zone to the adjacent abducens nucleus. This suggested that the high-angular resolution fiber tracking was able to distinguish the facial nerve from the vestibulocochlear nerve complex. The tractography clearly visualized cranial nerves displaced by brain tumors. These tractography findings were confirmed intraoperatively. Further studies involving larger sample sizes are needed to evaluate the clinical usefulness of the proposed technique [15].

18.3   Combined Techniques: Overcoming the Limitations

The DTT technique does not allow detailed 3D assessment of the cranial nerves/tumor complex, as the fibers are displayed as tridimensional bundles superimposed to 2D tumor scans. Chen et al. [4] assessed whether DTT of the cranial nerves combined with anatomic MRI of the tumor could provide superior three-dimensional (3D) visualization. DTI and anatomic images were analyzed in three subjects with VS; the anatomic images were used to model the 3D volume reconstruction of the tumor. The two sets of images were then superimposed through the use of linear registration. The two combined techniques were effectively proven to consistently reconstruct the 3D spatial relationship of cranial nerves/tumor complexes and allowed for superior visualization compared with two-dimensional imaging. This technique can be a useful adjunct in both radiosurgical planning and neuronavigation; in fact, in the case of radiosurgical treatment, this combination of techniques would allow contouring and measurement of radiation dose to the cranial nerves [6]. As regards current commercial neuronavigation software, it is not detailed enough to allow modeling of small regions of interest (ROIs) such as the cranial nerves of the CPA. However, the intraoperative real-time comparison between DTT NVII predicted course and surgical finding could be feasible after the adoption of a tractography-integrated neuronavigation system for the verification process [16].

Yoshino et al. [17] assessed whether the combined use of DTT and contrast-enhanced (CE) fast imaging employing steady-state acquisition (FIESTA) could improve the accuracy of predicting the courses of the facial and—most importantly—cochlear nerves before surgery, as the predictivity of DTT alone in relation to the cochlear nerve position has been described rarely and poorly before [9]. Twenty-two patients with VS were enrolled in the study, in whom both N VII and cochlear nerve could be identified during surgery. The rate of candidates for nerves predicted by combined DTT and multifused CE-FIESTA coinciding with the cochlear nerve was 63.6% (14/22 patients); however, that of candidates for nerves predicted by combined DTT and multifused CE-FIESTA coinciding with the facial nerve was 63.6% (14/22), much lower than the figures reported in the studies reviewed by Savardekar et al. [5].

Zolal et al. [18] did not probely experiment a combination of techniques featuring DTT; rather, they proposed that probabilistic non-tensor-based tractography might offer advantages in terms of better extraction of fiber directional information in areas where multiple fiber populations occupy the same voxel, as is the case of cranial nerves which are of sub-voxel size. In addition, probabilistic methods take into consideration the uncertainty of the data and model further possible directions in each step. In contrast to determinist tracking, probabilistic methods result in probability maps representing the likelihood of a voxel being part of the connection that is being sought after. Twenty-one patients with large vestibular schwannomas were recruited. The probabilistic tracking was run preoperatively, and the position of the potential depictions of the facial and cochlear nerves was estimated postoperatively; the true position of the nerves was determined intraoperatively by the surgeon. The probabilistic tracking showed a connection that correlated to the position of N VII in 81% of the cases and to the position of the cochlear nerve in 33% of the cases [18]. Larger studies are needed to further explore the potential advantages of probabilistic techniques.

References

1. Gerganov VM, Giordano M, Samii M, Samii A. Diffusion tensor imaging-based fiber tracking for prediction of the position of the facial nerve in relation to large vestibular schwannomas. J Neurosurg. 2011;115(6):1087–93.
2. Taoka T, Hirabayashi H, Nakagawa H, Sakamoto M, Myochin K, Hirohashi S, et al. Displacement of the facial nerve course by vestibular schwannoma: preoperative visualization using diffusion tensor tractography. J Magn Reson Imaging. 2006;24(5):1005–10.

3. Sartoretti-Schefer S, Kollias S, Valavanis A. Spatial relationship between vestibular schwannoma and facial nerve on three-dimensional T2-weighted fast spin-echo MR images. AJNR Am J Neuroradiol. 2000;21(5):810–6.

4. Chen DQ, Quan J, Guha A, Tymianski M, Mikulis D, Hodaie M. Three-dimensional in vivo modeling of vestibular schwannomas and surrounding cranial nerves with diffusion imaging tractography. Neurosurgery. 2011;68(4):1077–83.

5. Savardekar AR, Patra DP, Thakur JD, Narayan V, Mohammed N, Bollam P, et al. Preoperative diffusion tensor imaging-fiber tracking for facial nerve identification in vestibular schwannoma: a systematic review on its evolution and current status with a pooled data analysis of surgical concordance rates. Neurosurg Focus. 2018;44(3):E5.

6. Choi KS, Kim MS, Kwon HG, Jang SH, Kim OL. Preoperative identification of facial nerve in vestibular schwannomas surgery using diffusion tensor tractography. J Korean Neurosurg Soc. 2014;56(1):11–5.

7. Hodaie M, Quan J, Chen DQ. In vivo visualization of cranial nerve pathways in humans using diffusion-based tractography. Neurosurgery. 2010;66(4):788–95; discussion 95–6.

8. Zhang Y, Mao Z, Wei P, Jin Y, Ma L, Zhang J, et al. Preoperative prediction of location and shape of facial nerve in patients with large vestibular schwannomas using diffusion tensor imaging-based fiber tracking. World Neurosurg. 2017;99:70–8.

9. Wei PH, Qi ZG, Chen G, Hu P, Li MC, Liang JT, et al. Identification of cranial nerves near large vestibular schwannomas using superselective diffusion tensor tractography: experience with 23 cases. Acta Neurochir (Wien). 2015;157(7):1239–49.

10. Song F, Hou Y, Sun G, Chen X, Xu B, Huang JH, et al. In vivo visualization of the facial nerve in patients with acoustic neuroma using diffusion tensor imaging-based fiber tracking. J Neurosurg. 2016;125(4):787–94.

11. Hilly O, Chen JM, Birch J, Hwang E, Lin VY, Aviv RI, et al. Diffusion tensor imaging tractography of the facial nerve in patients with cerebellopontine angle tumors. Otol Neurotol. 2016;37(4):388–93. https://www.ncbi.nlm.nih.gov/pubmed/26905823.

12. Ung N, Mathur M, Chung LK, Cremer N, Pelargos P, Frew A, et al. A systematic analysis of the reliability of diffusion tensor imaging tractography for facial nerve imaging in patients with vestibular schwannoma. J Neurol Surg B Skull Base. 2016;77(4):314–8.

13. Dunn IF, Bi WL, Mukundan S, Delman BN, Parish J, Atkins T, et al. Congress of Neurological Surgeons systematic review and evidence-based guidelines on the role of imaging in the diagnosis and management of patients with vestibular schwannomas. Neurosurgery. 2018;82(2):E32–E4.

14. Roundy N, Delashaw JB, Cetas JS. Preoperative identification of the facial nerve in patients with large cerebellopontine angle tumors using high-density diffusion tensor imaging. J Neurosurg. 2012;116(4):697–702.

15. Yoshino M, Abhinav K, Yeh FC, Panesar S, Fernandes D, Pathak S, et al. Visualization of cranial nerves using high-definition fiber tractography. Neurosurgery. 2016;79(1):146–65.

16. Li H, Wang L, Hao S, Li D, Wu Z, Zhang L, et al. Identification of the facial nerve in relation to vestibular schwannoma using preoperative diffusion tensor tractography and intraoperative tractography-integrated neuronavigation system. World Neurosurg. 2017;107:669–77.

17. Yoshino M, Kin T, Ito A, Saito T, Nakagawa D, Ino K, et al. Combined use of diffusion tensor tractography and multifused contrast-enhanced FIESTA for predicting facial and cochlear nerve positions in relation to vestibular schwannoma. J Neurosurg. 2015;123(6):1480–8.

18. Zolal A, Juratli TA, Podlesek D, Rieger B, Kitzler HH, Linn J, et al. Probabilistic tractography of the cranial nerves in vestibular schwannoma. World Neurosurg. 2017;107:47–53.

Vestibular Testing to Predict the Nerve of Origin of Vestibular Schwannomas

19

Alberto Campione, Guglielmo Cacciotti,
Raffaelino Roperto, Carlo Giacobbo Scavo,
and Luciano Mastronardi

Vestibular schwannomas (VS) mainly arise from either the superior vestibular nerve (SVN) or inferior vestibular nerve (IVN). The SVN innervates the lateral semicircular canal (LSC) and anterior semicircular canal (ASC), utricle, and part of the saccule. On the other hand, the IVN innervates the posterior semicircular canal (PSC) as well as most of the saccule. On this basis, preoperative vestibular testing in patients affected by VS can be useful to predict which one of the vestibular nerves the tumor arises from. The relevance of identifying the nerve of origin in VS lies in its prognostic factor for hearing preservation after surgery [1–5], with tumors arising from the SVN having a 61–80% of hearing preservation rate, compared to 16–43% for an IVN origin [1, 4, 5] in cases when hearing preservation is attempted.

Diverse studies in the literature have tested the correlation between asymmetric or pathological vestibular testing results and the nerve of origin of VSs, with subsequent controversial conclusions. The techniques that have been experimented in such context are posturography, vestibular evoked myogenic potentials (VEMPs), caloric test (always in combination with VEMPs), and video head impulse test (vHIT).

19.1 Posturography

Computerized dynamic platform posturography (CDPP) is a sensory organization test that consists of six conditions (i.e., the steps of the examination) of increasing difficulty in separate 20 s trials; a balance score ranging from 0% (worst) to 100% (best) is assigned for each condition. Conditions 5 and 6 assess the vestibular component of the balance system separately by eliminating, through sway-referencing, information from vision and somatosensation, respectively. During condition 5, the

A. Campione (✉) · G. Cacciotti · R. Roperto · C. Giacobbo Scavo · L. Mastronardi
Department of Neurosurgery, San Filippo Neri Hospital—ASLRoma1, Rome, Italy
e-mail: mastro@tin.it

© Springer Nature Switzerland AG 2019
L. Mastronardi et al. (eds.), *Advances in Vestibular Schwannoma Microneurosurgery*, https://doi.org/10.1007/978-3-030-03167-1_19

patient stands on the mobile, sway-referenced platform with their eyes closed. During condition 6, the patient stands with their eyes open on a mobile, sway-referenced platform with a sway-referenced visual surround. Condition 5 score (C5S) and condition 6 score (C6S) are the respective arithmetic means (in %) of the scores recorded during three repetitions of each condition. Results below the fifth percentiles of age-matched normal individuals are rated as pathologic [1].

Gouveris et al. [6] performed a retrospective study to test whether CDPP findings could preoperatively predict the nerve of origin of VSs. Seventy-five consecutive VS patients were evaluated; C5S, C6S, vestibular ratio (VER), and mean overall balance score (MOBS) were calculated for each patient. The nerve of VS origin was identified intraoperatively. Although lower median values for C5S and C6S were observed in patients with SVN compared with IVN tumors, none of the four scores showed any significant difference between the SVN and IVN groups of VS patients [6].

Borgmann et al. [1] combined posturography and caloric electronystagmography (ENG); caloric testing was done by standard bithermal irrigation of the vestibular organ, using water at 30 and 44 °C, to stimulate the LSC and thus yield a functional depiction of SVN functional state. Eye movements were recorded by ENG, and the maximal slow-phase eye velocity was used to calculate canal paresis. A percentage of left-right difference $\geq 25\%$ was defined as pathologic. Eighty-nine patients with VS originating from the IVN and 22 patients with VS from the SVN were included. Pathologic results in preoperative caloric ENG ($p < 0.0001$) and CDPP ($p = 0.025$) were significantly more frequent in subjects with SVN than with IVN VSs. In addition, hearing preservation rate was significantly higher in patients with tumors from the SVN than from the IVN ($p = 0.011$) [1].

19.2   VEMPs and Caloric Test

VEMPs' apparatus arrangement may differ according to the protocols used; however, a surface electrode is always placed on the upper half of the sternocleidomastoid muscle ipsilateral to the stimulated ear along with a reference electrode on the upper sternum and a ground electrode on the nasion [4, 7]. The patients are then instructed to rotate their heads toward the non-stimulated ear side. During recording, electromyographic activities are monitored on a display to maintain muscle activity at a constant level. Clicks and bursts (of different duration and intensity according to the chosen protocol) are presented through a headphone at a determinate stimulation rate. After response averaging, the amplitude of the first positive-negative peak—i.e., p13-n23—is analyzed, and the percentage of response asymmetry between the two sides is calculated, if present. The responses of VEMPs are regarded as abnormal when the responses on the affected side are absent or decreased compared with those of the unaffected side [4, 7].

VEMPs and caloric test can be regarded as complementary vestibular tests. In fact, they allow for IVN and SVN individual examination, respectively. VEMPs are designed to elicit the vestibulocollic reflex, thus stimulating ASC and PSC, the latter of which is innervated by IVN. On the other hand, the injection of either cold [4, 7, Ushio] or hot [1] water into the external auditory meatus evokes the vestibulo-ocular reflex (VOR) (recorded by ENG), thus stimulating the LSC, innervated by SVN.

Tsutsumi et al. [8] performed a retrospective study to determine whether the nerve of origin of VSs could be predicted using VEMPs alone. Twenty-eight patients undergoing surgical resection were included in the analysis. Complete disappearance of VEMPs was observed only in patients with tumors arising from IVN; therefore the authors concluded that prediction of the nerve of origin was possible only in certain restricted cases [8].

Ushio et al. [9] described a series of 109 consecutive patients diagnosed as having unilateral VS; each of them underwent both VEMPs and caloric test evaluation before surgery. The nerve of origin of the tumor was identified in 63 of the 109 patients. The percentage of patients showing abnormal responses in each test was not different between 37 patients with SVN VSs and 26 patients with IVN VSs: abnormal caloric responses were seen in 86.5% (32/37) of patients with superior VS and in 80.8% (21/26) of patients with inferior VS ($p = 0.54$), and abnormal VEMPs responses were shown in 77.4% (24/31) of patients with superior VS and in 66.7% (12/18) of patients with inferior VS ($p = 0.41$) [9].

Suzuki et al. [7] reported results similar to those of the study by Ushio et al. In their 130-patient series, abnormal caloric and VEMPs response rates in patients with tumors arising from the SVN were not significantly different from those in patients with tumors of the IVN ($\chi^2 = 0.618$ for caloric test responses, SVN vs. IVN tumor; $\chi^2 = 0.715$ for VEMPs responses, SVN vs. IVN tumors) [7].

In the prospective study by Chen et al. [10], eight patients with a cerebellopontine angle (CPA) tumor underwent caloric test and VEMPs examination. Four of the eight patients received surgical intervention, which included three cases of VS and one epidermoid cyst. Follow-up study was performed 1 year after the surgery. During the surgery, the patients with neither caloric response nor VEMPs had a tumor that involved both SVN and IVN. On the contrary, in the one patient with a normal caloric response and absent VEMPs, the tumor originated from IVN. In the follow-up caloric test and VEMPs examination, only one patient with an epidermoid cyst had a complete recovery in both tests, whereas the other three VS patients with absent VEMPs were unchanged. In spite of the limited number of cases, the authors concluded that before surgery, VEMPs test could be used to predict the nerve of origin and to formulate the best surgical approach. After surgery, VEMPs test could be used to define the nature of the tumor (compressing or infiltrating the nerve) and disclose the residual function of the IVN [10].

He et al. [4] conducted as well a prospective study and enrolled 106 VS patients, who received both caloric test and VEMPs examination before the surgical procedure and during follow-up. During the operation, the nerve of origin (SVN or IVN) was identified by the surgeon; proper identification of the nerve of origin was feasible in 68 patients. The tumors arose from the SVN in 26 patients and from the IVN in 42 patients. The results of the caloric tests and VEMPs tests were significantly different in tumors originating from SVN and IVN. The combination of abnormal VEMPs and normal caloric test response yielded a positive predictive value (PPV) as high as 21.4% of tumor originating from IVN; on the other hand, the combination of normal VEMPs and abnormal caloric tests yielded a PPV as high as 50% of tumor originating from SVN. The authors concluded that caloric and VEMPs tests might help to identify whether VSs originate from the SVN or IVN and that such tests could also be used to evaluate the residual function of the nerves after surgery [4].

19.3   Video Head Impulse Test

The vHIT is a noninvasive test that allows quantitative evaluation of the gain of VOR as well as the identification of covert (occurring while the head is still moving) and overt (occurring once the head movement is finished) saccades on LSCs, ASCs, and PSCs. The procedure consists in the analysis of eye movement (with a video-oculography camera) during head movement, detected and quantified by a dedicated sensor. The patient is asked to fix his gaze at a target 1 m away; then, the examiner rotates the patient's head randomly 15°–20° on the horizontal plane, thus allowing for the evaluation of both LSCs. The vertical semicircular canals are evaluated with a 45° head rotation to the right (left ASC and right PSC) and to the left (right ASC and left PSC), each followed by an anterior and then a posterior impulse. Twenty stimuli for every semicircular canal are performed to assure a sustained response. The evaluated parameters are gain of VOR (relationship between the velocity of head and eye movements) of every canal, expressed as percentage to evaluate the functional deficit of the affected ear, and the presence of overt and covert saccades. Gain of VOR is categorized as normal or abnormal according to age-dependent normative values. Refixation saccades (both covert and overt) are a physiological phenomenon used by the central vestibular pathways to compensate for the low gain of VOR—thus, they are a sign of a deficit in VOR efficiency [11]. As vHIT analyzes each of the semicircular canals singularly, the detected alterations may be interpreted as indirect signs of pathological compression or infiltration of the corresponding vestibular nerve.

Rahne et al. [12] conducted a study to introduce a novel scoring system that was designed to determine the nerve of origin of VSs, based on vHIT and cervical/ocular VEMPs. The rationale in collecting results of both the functional tests was to gain data as complete as possible about the functional state of both SVN and IVN. In fact, vHIT analyzes the semicircular canals but not the utricle or the saccule. Instead, cervical VEMPs correlate with saccular function and IVN activity as well as ocular VEMPs correlate with utricular function and SVN activity. The parameters included in the scoring system were abnormal gain of VOR and presence of saccades for each of the semicircular canals, abnormal cervical VEMPs, and abnormal ocular VEMPs. The preoperatively acquired data were entered into the scoring system, and the nerve of tumor origin was eventually determined intraoperatively. The scoring system was applied to five consecutive patients undergoing surgical VS treatment. In one case, no determination was possible—this was the largest tumor of the cohort, Koos Grade IV. In all the other cases, the preoperatively predicted tumor origin corresponded to the surgical finding, so that the experimented scoring system yielded a PPV of 100% [12].

Costanzo et al. [11] preoperatively evaluated 31 VS patients with vHIT (gain of VOR, overt and covert saccades on each semicircular canal were reported); the nerve of origin was identified intraoperatively during surgical resection. Surgical identification of the nerve of origin was achieved in 29 of the 31 patients, both the remaining cases being Hannover-T4b lesions. Of the 19 surgically identified SVN schwannomas, vHIT showed a SVN dysfunction pattern in 17 cases and a normal response in 2 cases, giving a correct preoperative diagnosis rate—i.e., a PPV—of 89.5%. Of the ten IVN lesions, vHIT showed an IVN dysfunction pattern in nine cases and a normal test in one case, giving a correct preoperative diagnosis rate of 81.8%. Overall, vHIT lead to a correct identification of the nerve of origin in 100% of altered exams and in 26 (89.7%) of the 29 surgically identified cases. Therefore, the authors concluded that the pattern of semicircular canal dysfunction on vHIT has a localizing value to identify the nerve of origin in VSs [11].

References

1. Borgmann H, Lenarz T, Lenarz M. Preoperative prediction of vestibular schwannoma's nerve of origin with posturography and electronystagmography. Acta Otolaryngol. 2011;131(5):498–503.
2. Brackmann DE, Owens RM, Friedman RA, Hitselberger WE, De la Cruz A, House JW, et al. Prognostic factors for hearing preservation in vestibular schwannoma surgery. Am J Otol. 2000;21(3):417–24.
3. Cohen NL, Lewis WS, Ransohoff J. Hearing preservation in cerebellopontine angle tumor surgery: the NYU experience 1974-1991. Am J Otol. 1993;14(5):423–33.
4. He YB, Yu CJ, Ji HM, Qu YM, Chen N. Significance of vestibular testing on distinguishing the nerve of origin for vestibular schwannoma and predicting the preservation of hearing. Chin Med J (Engl). 2016;129(7):799–803.
5. Jacob A, Robinson LL, Bortman JS, Yu L, Dodson EE, Welling DB. Nerve of origin, tumor size, hearing preservation, and facial nerve outcomes in 359 vestibular schwannoma resections at a tertiary care academic center. Laryngoscope. 2007;117(12):2087–92.
6. Gouveris H, Akkafa S, Lippold R, Mann W. Influence of nerve of origin and tumor size of vestibular schwannoma on dynamic posturography findings. Acta Otolaryngol. 2006;126(12):1281–5.
7. Suzuki M, Yamada C, Inoue R, Kashio A, Saito Y, Nakanishi W. Analysis of vestibular testing in patients with vestibular schwannoma based on the nerve of origin, the localization, and the size of the tumor. Otol Neurotol. 2008;29(7):1029–33.
8. Tsutsumi T, Tsunoda A, Noguchi Y, Komatsuzaki A. Prediction of the nerves of origin of vestibular schwannomas with vestibular evoked myogenic potentials. Am J Otol. 2000;21(5):712–5.
9. Ushio M, Iwasaki S, Chihara Y, Kawahara N, Morita A, Saito N, et al. Is the nerve origin of the vestibular schwannoma correlated with vestibular evoked myogenic potential, caloric test, and auditory brainstem response? Acta Otolaryngol. 2009;129(10):1095–100.
10. Chen CW, Young YH, Tseng HM. Preoperative versus postoperative role of vestibular-evoked myogenic potentials in cerebellopontine angle tumor. Laryngoscope. 2002;112(2):267–71.
11. Constanzo F, Sens P, Teixeira BC de A, Ramina R. Video head impulse test to preoperatively identify the nerve of origin of vestibular schwannomas. Oper Neurosurg. 2018. https://doi.org/10.1093/ons/opy103.
12. Rahne T, Plößl S, Plontke SK, Strauss C. Preoperative determination of nerve of origin in patients with vestibular schwannoma. German version. HNO. 2017;65(12):966–72.

Microsurgery for Vestibular Schwannomas After Failed Radiation Treatment

20

Yoichi Nonaka and Takanori Fukushima

Abbreviations

CK	CyberKnife
CN	Cranial nerve
GTR	Gloss total resection
H-B	House-Brackmann facial nerve function grading scale
LINAC	Linear accelerator
MR-VS	Vestibular schwannoma previously treated with microsurgery and radiotherapy
NTR	Near-total resection
N-VS	Non-radiated vestibular schwannoma
R-VS	Radiated vestibular schwannoma
SRS	Stereotactic radiosurgery
SRT	Stereotactic radiotherapy
STR	Subtotal resection
VS	Vestibular schwannoma

Currently, there are three primary management strategies for small- to medium-sized vestibular schwannomas (VSs), including wait-and-scan, microsurgical resection, and radiation therapy. Numbers of patients with VS treated with stereotactic radiation therapy (SRT) have been increasing over the past two decades with the

Y. Nonaka (✉)
Department of Neurosurgery, Tokai University School of Medicine, Kanagawa, Japan

T. Fukushima
Division of Neurosurgery, Duke University Medical Center, Carolina Neuroscience Institute, Raleigh, NC, USA
e-mail: Fukushima@carolinaneuroscience.com

© Springer Nature Switzerland AG 2019
L. Mastronardi et al. (eds.), *Advances in Vestibular Schwannoma Microneurosurgery*, https://doi.org/10.1007/978-3-030-03167-1_20

trend of less use of microsurgery. Several types of focused-beam SRT including Gamma Knife (GK), CyberKnife (CK), Novaris, and proton beam have been utilized to control the growth of VS since the early 1990s. With advances in computer technology and more accurate targeting, SRT has become widely adopted to the treatment option of VS. Despite these advances, a small percentage of VS grow after radiation treatment. However, management for these patients is still controversial. Although, repeat radiation therapy is thought to carry increased adverse risk and a higher rate of secondary failure, some of patients underwent two or more times of radiation therapy. On the other hand, surgical outcomes for VSs have improved following the progress in microsurgical techniques. In 2016, we reported our experience of surgical management for previously radiated VSs who required surgical intervention [1]. Our findings suggested that salvage surgery for VS who fail SRT would have a higher risk of postoperative complications. In this report, we updated our surgical experience with a consecutive series of 74 cases of VSs after failed radiation treatment.

20.1 Clinical Material and Methods

Between January 1995 and December 2016, 2115 patients with unilateral VSs were surgically treated by the authors. Neurofibromatosis type II patients were excluded from this study. Seventy-four patients (3.5%) were identified who had previous SRT elsewhere one or more times prior to surgery. Out of 74 patients, 48 had been treated by radiation therapy only (R-VS), and 26 had a combination of microsurgical tumor debulking and radiation therapy (MR-VS). There were 24 males and 50 females. Age ranged from 14 to 73 years (mean age 51.8). Fifty-five patients (74.3%) were treated using GK, six (8.1%) were treated using fractionated stereotactic radiotherapy, four (5.4%) were treated using CK, and one (1.4%) was treated using proton beam. Unknown type of radiation therapy had done for seven patients (9.5%). Four patients had SRT more than twice. Only one patient had treatment with CK after failed GK. In all cases, initial radiosurgical procedures were performed at various neurosurgical centers.

Patient characteristics are summarized in Table 20.1. Preoperative neurological findings in R-VS were analyzed to document symptoms that worsened or first developed after SRT. Facial nerve function was evaluated according to the House-Brackmann facial nerve function grading scale (H-B grade) [2]. Tumor size was measured as maximal extrameatal tumor diameter on the postcontrast axial MRI according to the International Criteria for Vestibular Schwannoma reported by Kanzaki et al. [3]. If the tumor had cyst formation, the size of cyst was also included in the tumor size. Due to a lack of precise information regarding the treatment and its indications, we were not able to pursue this issue further. Operative reports and video recordings were reviewed for the tumor characteristics, severity of fibrous adhesions, any unusual appearance, and relationship of tumor capsule with neurovascular structures.

Table 20.1 Characteristics in 74 patients with irradiated VSs

Characteristics			Value (%)
Age (years)			
Range			14–73
Average			51.8
Sex			
Male			24 (32.4)
Female			50 (67.6)
Previous treatment			
Radiation therapy only			48 (64.9)
Previous surgery			26 (35.1)
Size of tumors at salvage surgery (mm)			
Intracanal		0	(0)
Small	(1–10)	5	(6.8)
Medium	(11–20)	12	(16.2)
Moderately large	(21–30)	32	(43.2)
Large	(31–40)	19	(25.7)
Giant	(≥41)	6	(8.1)
Kind of SRT			
GK			55 (74.3)
Fractionated stereotactic radiotherapy			6 (8.1)
CK			4 (5.4)
Proton beam			1 (1.4)
GK + CK			1
Unknown			7 (9.5)
Interval between SRT and surgery (month)			45.1 (range 8–240)
Surgical approaches			
RS approach			52 (70.3)
TL approach			22 (29.7)
Extent of tumor resection			
GTR			25 (37.8)
NTR			14 (31.1)
STR			14 (31.1)

CK CyberKnife, *GK* gamma knife, *GTR* gross total resection, *LINAC* linear accelerator, *NTR* near-total resection, *RS* retrosigmoid, *SRT* stereotactic radiation treatment, *STR* subtotal resection, *TL* translabyrinthine

Postoperative outcomes of R-VSs were compared with those from our consecutive series of 379 cases of non-radiated VSs (N-VS) between 2000 and 2009 previously reported [4]. Surgery is basically indicated for patients with definitive tumor growth after more than 3 years following SRT. However, patients with rapid tumor growth and worsening of neurological symptoms due to mass effect underwent tumor resection prior to 3 years.

Extent of tumor resection was graded into three categories. Category 1: Gross total resection (GTR) means a total tumor removal by the surgeon's determination, with no residual tumor detected on postoperative contrast MRI. Category 2: Near-total resection (NTR) means a small trace (<0.5 mm) of the tumor capsule remains on the thinned and stretched CN VII and VIII or on the brainstem. Postoperative MRI shows a thin line of enhancement (<1–2% of original mass). Category 3:

Subtotal resection (STR) means a few millimeters thickness of tumor capsule is left with CN VII and VIII or on the brainstem. Postoperative MRI shows a residual tumor capsule, approximately 5–10% of the original volume.

20.2 Results

Fifty-nine patients (79.7%) demonstrated steady tumor growth after SRT, while 15 patients (20.3%) had a rapid tumor growth after several years of quiescence. Seven patients (9.5%) without definitive tumor growth underwent surgical resection for unbearable facial pain, psychological distress, or patient's request. Tumor size at the time of salvage surgery is depicted in Table 20.1. The interval between SRT and surgery ranged 8–240 months (mean 45.1 months). GTR, NTR, and STR were achieved in 25 (37.8%), 14 (31.1%), and 14 (31.1%) patients, respectively. Symptoms worsened or newly developed following SRT consisted of dizziness (35.9%), ataxia or disequilibrium (33.3%), and tinnitus in 20.5% of patients. Hearing decrease was noted in 15.4% and deafness was observed in 41% following SRT. Worsening of trigeminal facial pain was observed in 7.7% of patients. Facial numbness was observed in 25.6% and facial palsy was detected in 7.7%. Devastating ataxia was observed in one patient. Lower cranial nerve deficits were seen in 5.1%. All of these symptoms first developed or significantly worsened immediately following SRT. A comparison of intraoperative videos for R-VS to our large series of N-VS cases identified certain differences attributable to the radiation effect.

Arachnoid membrane around tumor was thickened and opaque. Radiated tumors exhibited unusual fibrous change and tenacious contents of the tumor tissue in 46.2% of cases. Newly cyst formation was seen in 23.1% and brownish/purple discoloration of the tumor capsule in 15.4% cases. Additionally, severe fibrous adhesions between the tumor capsule and CNs, vessels, or brainstem were observed in 69.2%. In larger tumors, the facial nerve and the brainstem surface appeared much softer and more fragile in 17.9% of cases suggesting radiation-influenced neuromalacia. This was further supported by the increased frequency of injury potentials in the facial nerve monitor in response to even very mild manipulation.

There were neither mortality nor any major complications in this series. Preexisting facial nerve palsy or weakness was seen in 5 out of 74 patients (6.8%) at the time of surgery. These patients had no worsening of facial weakness following surgery. Twelve out of 69 patients (17.3%) newly developed facial nerve palsy postoperatively. H-B grade III was seen in eight patients, grade IV was in five patients, and one patient had grade V facial nerve palsy.

20.3 Illustrative Cases

A 58-year-old woman presented with 3-year history of right-sided hearing loss, tinnitus, and increasing dizziness. Initial MRI demonstrated a right-sided intracanalicular tumor with a few millimeters of protrusion into the cerebellopontine angle

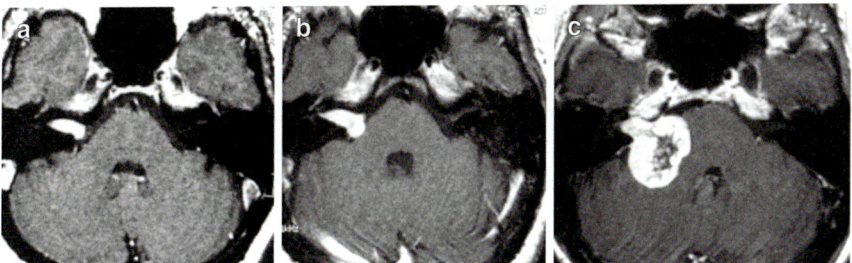

Fig. 20.1 Illustrative case 1: A 58-year-old female. (**a**) A postcontrast axial MRI reveals right intrameatal tumor with homogeneous enhancement (before GK treatment). (**b**) A postcontrast axial MRI demonstrates slight growth of extrameatal portion (2 years after GK treatment). (**c**) A postcontrast axial MRI demonstrates further tumor growth compressing the brainstem and the cerebellum (3 years after GK treatment). The tumor lacks central enhancement

(Fig. 20.1a). The patient underwent GK radiation with margin dose 12.3 Gy and was followed with annual MR imaging. A slight tumor enlargement of the extrameatal portion was seen 2 years after the radiation treatment (Fig. 20.1b). During the next follow-up period, the patient developed total hearing loss in the treated ear, worsening dizziness, and disequilibrium. An MRI taken at 3 years post SRT revealed a substantial regrowth of the tumor in a year with compression of the brainstem (Fig. 20.1c). After 3 years of observation by a radiation specialist, this patient was referred to our service for surgery.

20.4 Discussion

Over the past two decades, increasing numbers of patients with VS have been treated with SRT [5–19]. According to the reported series, the incidence of tumor growth with SRT has been favorable [6, 9, 12, 16, 19, 20]. Long-term follow-up outcomes of radiosurgery for small- to medium-sized VSs from recent papers revealed that tumor regrowth control in 92–97% of patients up to 10 years following GKS [8, 12]. However, with the exception of mild tumor shrinkage in some cases, patients who undergo SRT for VS will have persistent tumor mass and will require continued close surveillance to rule out future regrowth of the tumor and worsening of neurological symptoms. On the other hand, a majority of patients who undergo microsurgery by experts will have total or near-total resection of the tumor for persistent cure. As compared with prior SRT reports, our series demonstrated a large number of symptoms that started or worsened immediately after SRT such as deafness, facial numbness or hypoesthesia, facial nerve palsy, and facial pain [17, 21–23]. It is important to discuss these findings when counseling patients who are considering SRT. Although it is quite rare, another fact that must be mentioned is that malignant transformation of VS by SRT has been documented in 38 patients [24–33].

Temporary tumor expansion within 6–12 months after SRT is a well-known phenomenon occurring in 2–45% of patients [23, 34]. Pollock et al. reported that the median time to tumor enlargement after stereotactic radiosurgery was 9 months, and the median volume increase was 75% [23]. In their series, new cranial neuropathies occurred during this period, but many were temporary and resolved without treatment. Therefore, it is our general practice to observe these patients for at least 3 years following their SRT, unless they develop severe symptoms requiring earlier surgical intervention. Consequently, the mean interval between SRT and salvage surgery in the reported literature was 32.1 months (range 19.2–46 months) and 45.1 months (range 8–240 months) in the present series [34–48].

Less than 10% of radiation failure with progressive tumor growth was reported. Surgical resection after radiosurgery is indicated when symptoms as cerebellar ataxia, increased intracranial pressure, or progressive symptoms occur at the time of tumor growth, even if this is during the time of transient expansion. In our previous study of 39 patients who needed surgery after onetime or multiple rounds of SRT for VS, we found that the number of previously radiated patients requiring surgery accounted for an increasing fraction of our VS surgery practice. These patients required surgery for a variety of reasons. We had only 2 patients representing 0.8% of our total VS population in 1995–1999, 8 patients or 2.7% of our total practice in 2000–2005, and 64 patients or 8.9% of our total practice in 2006–2013. This trend may be a consequence of the more widespread and perhaps inappropriate use of this treatment modality. Alternatively, it may represent the accumulation of treatment failures as the years of follow-up increases. While some of these tumors continue to grow at a steady rate following SRT, approximately 10% of these tumors were quiescent for many years before starting to grow rapidly. This highlights the perils of limited follow-up duration seen in some of the literature. Benign tumors such as VS require at least 15–20 years of follow-up for definitive treatment outcomes to be determined. Current studies of long-term outcomes seem to indicate that SRT may be a promising and valuable technology [8, 11, 16, 49]. However, the observation period is still too short to determine its efficacy for such slow-growing tumors.

Most authors agree that operating on previously radiated VS is more difficult compared to primary treatment [34–38, 40–48]. Facial nerve dissection was noted to be subjectively more difficult secondary to adherent and/or poorly defined surgical planes at the nerve-tumor interface. Reirradiation carries an increased risk of cranial neuropathy, hydrocephalus, and radiation-induced cerebral edema or necrosis. Furthermore, tumors that initially fail radiation therapy may be considered more radioresistant and less likely to respond to additionally radiation treatment. Considering these facts, the authors favor microsurgical salvage in most cases [1]. In 2012, Gerganov et al. also found a much higher rate of poor facial nerve function after salvage surgery in patients who had prior radiation therapy [38]. Wise et al. similarly reviewed a series of 37 patients who failed primary SRS [48]. They presented a large multicenter case-control study comparing outcomes between postradiated sporadic VS and non-radiated control subjects. Approximately 77% of patients with normal preoperative facial nerve function retained good function after salvage surgery, and GTR and NTR were achieved in 49% and 27% of the cases, respectively. Due to the increased experience of surgery after radiosurgery, the surgical strategy is becoming more conservative for functional preservation.

We also found a variety of postradiation changes depicted in Fig. 20.2. Adhesions between the tumor capsule and CNs, brainstem, or vessels were most frequently seen. While these are also seen in N-VS, the adhesions are thicker, more gluelike, and harder to dissect in radiation-failed VS (Fig. 20.2a, b). In some cases, a dissection plane could not be established between the tumor and the nerve (Fig. 20.2c). Furthermore, the arachnoid membrane around the tumor was thicker, more opaque, and stickier than non-irradiated tumors (Fig. 20.2d). Neuromalacia or softening of

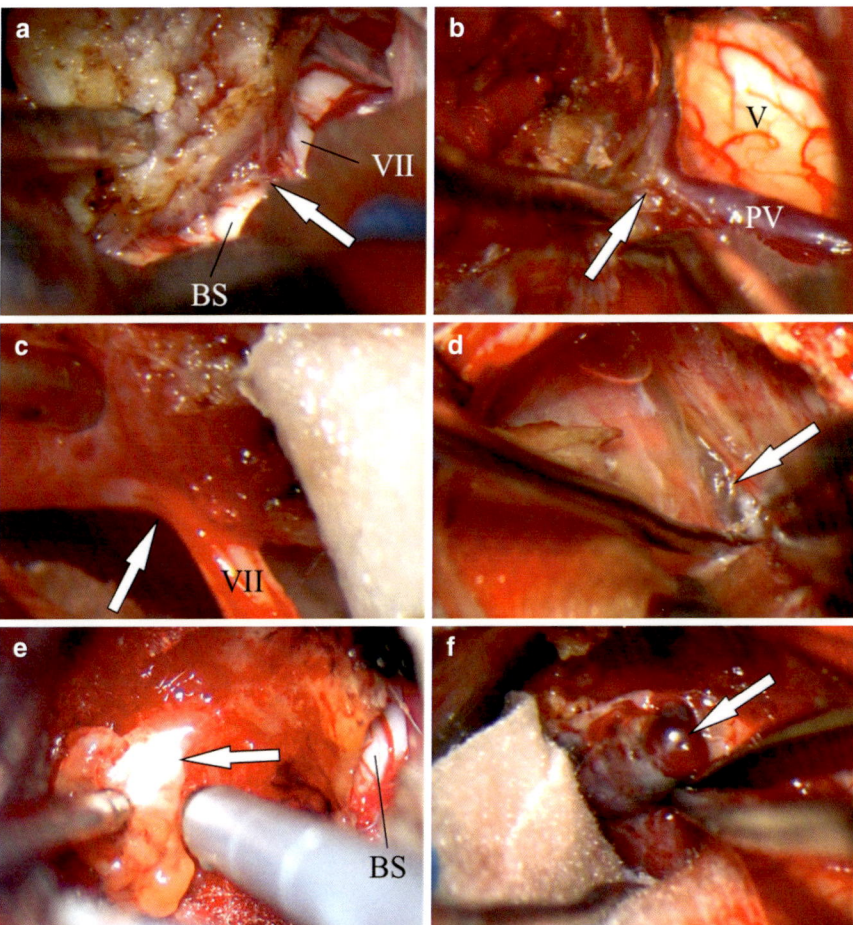

Fig. 20.2 Intraoperative video still images demonstrate peculiar features of irradiated VS. (**a**) (R-RS) severe adhesion between the tumor capsule and surface of the brainstem (arrow), (**b**) (L-RS) adhesion between tumor capsule and petrosal vein (arrow), (**c**) (L-RS) facial nerve fuse into the tumor (arrow), (**d**) (L-RS) thicken arachnoid membrane between the tumor and tentorium (arrow), (**e**) (R-RS) sticky change of the tumor and fibrous change of central non-enhancement core of the tumor (arrow), (**f**) (R-RS) purple change of the tumor surface (arrow); *BS* brainstem, *VII* facial nerve, *PV* petrosal vein, *V* trigeminal nerve, *R-RS* right retrosigmoid approach, *L-RS* left retrosigmoid approach

the cranial nerves was particularly troublesome, as it could not be safely dissected from the tumor. The cerebellum was edematous and its surface was fragile. The tumor itself had some changes which complicated surgery as well: its surface was hardened with yellowish change, purple change was also seen due to subcapsular hemorrhages (Fig. 20.2e), and the tumor itself had both hard and tenacious parts. Fibrous change was seen at the central core of the tumor, corresponding to the lack of enhancement on the postcontrast MRI (Fig. 20.2f). These postradiation changes were not seen in all cases, but more often in cases with a long interval between radiation and surgical treatment. They are not detected on postirradiation MRI with the exception of brain edema and cyst formation. These are the surgeon's subjective observation; however, in the senior author's experience of over 1800 surgical cases of VS, this type of finding is extremely rare to see in non-radiated tumors. Similar findings have also been reported elsewhere in the literature (Table 20.2) [5, 34, 36–38, 40–45, 47, 50]. Other authors have commented on postradiation changes on

Table 20.2 Surgical outcomes for radiation-failed vestibular schwannomas in the literature

Author/year	No. of patients	Mean interval for surgery (months)	FN palsy[a] (%)	GTR (%)	Technical difficulties or tumor characteristics (subjective)
Slattery and Brackmann 1995	5	46	80	100	Tumors were severely scarred to FN
Pollock et al. 1998	13	27	61.5	53.8	Tumor fibrosis/loss of peritumoral arachnoidal plane
Battista and Wiet 2000	12	35	N/A	N/A	An absence of a tissue plane between tumor and FN (79%)
Lee et al. 2003	4	19.2 (1.6 years)	25	N/A	Dense adhesions and fibrosis were found along the cranial nerve
Friedman et al. 2005	38	39.6	57	78.9	Moderate to severe adherence of the tumor to the FN
Limb et al. 2005	8	N/A	62.5	N/A	Fibrous, scarring, and adherence of structures to the tumor
Iwai et al. 2007	6	28	33.3	0	Arachnoid thickening/intratumoral bleeding
Shuto et al. 2008	12	29	25	0	FN severe adhesion to the tumor/color change of the FN
Slattery 2009	62	37.2	N/A	79	Normal plane between the FN and tumor was difficult to determine
Lee et al. 2010	7	26	N/A	0	Authors did not think radiosurgical tumors were more difficult
Friedman et al. 2011	73	43.2 (3.6 years)	50[b], 14.3[c]	79.5	83.6% of tumors had moderate or severe adherence to the FN
Gerganov et al. 2012	28	30.7	20[d], 23.1[e]	100[d], 100[e]	More difficult due to the extensive arachnoid scarring
Hong et al. 2013	5[f]	N/A	33.3	60	Perceived to be more difficult/scarring and fibrosis of the arachnoidal layers

Table 20.2 (continued)

Author/year	No. of patients	Mean interval for surgery (months)	FN palsy[a] (%)	GTR (%)	Technical difficulties or tumor characteristics (subjective)
Wise et al. 2016	37	36	27	49	Subjectively more difficult secondary to adherent and or poorly defined surgical planes
Iwai et al. 2016	18	26	22	0	Difficulty in identifying the FN, probably due to radiation effect/ thickening of the arachnoid
Breshears et al. 2017	10	36	20	70	Noted to have significant or dense adhesions of the tumor to the BS or FN
Present series	74	45.6	17.3	37.8	Severe adhesion to the FN and BS, neuromalacia, lack of arachnoid plane, color change of the FN

BS brainstem, *FN* facial nerve, *GTR* gross total resection, *N/A* not available, *PR* partial resection, *STR* subtotal resection

[a]House and Brackmann grading system from IV, V, and VI was considered as facial nerve palsy (surgery related)

[b]Percentage in the group of GTR cases

[c]Percentage in the group of PR cases

[d]Group A (radiosurgery prior to surgery)

[e]Group B (partial tumor removal followed by radiosurgery prior to current surgery)

[f]Numbers of previously radiated patient

the neurovascular structures around the tumor [2, 6, 8, 11, 12, 25, 27, 34, 44, 51–53]. It is unknown whether these changes occur in SRT-controlled VSs; however, even if lower radiation doses (<10 Gy) are used, similar tissue effects would be expected. The irradiated CN, especially facial nerve, does not recover from microsurgical trauma when compared with the non-irradiated one [8, 53].

Facial nerve outcomes in R-VSs group were not worse than that of N-VSs because our team has been extremely careful for determination of the dissectability of the radiated facial nerve in our previous series [1]. In NTR or STR, some portion of the tumor capsule was left attached to the thin, adherent facial nerve or on the brainstem with the expectation of better facial nerve function. It stands to reason that leaving more residual tumor would translate into less FN injury. In the present series, we observed that the radiated facial nerve was softened, more fragile, and more densely adherent to the tumor capsule. Additionally, facial nerve monitors demonstrated more injury potentials under circumstances of minor manipulation than seen in N-VSs. Therefore the surgeon decided to stop dissecting tumor from the nerve after observing a decrease in facial nerve response. This may explain the higher than expected rate of FN palsy in the R-VSs group.

As many authors documented, GTR is not always feasible because of the severe fibrous adhesion of the tumor capsule with the brainstem and the thin facial nerve that is not resistant to microdissection [34, 36, 38–40, 46, 48]. To avoid complications, we recommend that NTR and STR should be performed to leave a thin capsule. It is not clear what the long-term results are for NTR or STR, but this is the recommended technique for minimizing postoperative morbidity in R-VS patients.

20.5   Conclusions

Microsurgical resection of VS following radiation is more difficult due to more fibrous adhesion, fibrosis of tumor capsule, and neuromalacia of the facial nerve when compared with our N-VS patients. These features are not apparent on preoperative imaging, but recognized difficulty during dissection to consider a planned subtotal resection, which may contribute to higher rate of facial nerve preservation. When previous SRT has failed, much longer follow-up is necessary to see the behavior of tumor remnants following NTR and STR. For now, the possibility of definitive tumor regrowth, the progression of clinical symptoms, the difficulty of operation with a decreased rate of gross total resection, and an increasing rate of facial nerve palsy following radiation must be discussed with patients considering SRT. We must be prepared for these difficult cases, and patients undergoing radiation for VS should be followed long term.

Disclosure   The authors report no conflict of interest concerning the materials or methods used in this study or the findings specified in this paper.

References

1. Nonaka Y, Fukushima T, Watanabe K, Friedman AH, Cunningham CD III, Zomorodi AR. Surgical management of vestibular schwannomas after failed radiation treatment. Neurosurg Rev. 2016;39:303–12.
2. House JW, Brackmann DE. Facial nerve grading system. Otolaryngol Head Neck Surg. 1985;93:146–7.
3. Kanzaki J, Tos M, Sanna M, Moffat DA, Monsell EM, Berliner KI. New and modified reporting systems from the consensus meeting on systems for reporting results in vestibular schwannoma. Otol Neurotol. 2003;24:642–8.
4. Nonaka Y, Fukushima T, Watanabe K, Friedman A, Sampson J, McElveen J, Cunningham C, Zomorodi A. Contemporary surgical management of vestibular schwannomas: analysis of complications and lessons learned over the past decade. Neurosurgery. 2013;72:ons103–15.
5. Battista RA, Wiet RJ. Stereotactic radiosurgery for acoustic neuromas: a survey of the American Neurotology Society. Am J Otol. 2000;21:371–81.
6. Chan AW, Black PM, Ojemann RG, et al. Stereotactic radiotherapy for vestibular schwannomas: favorable outcome with minimal toxicity. Neurosurgery. 2005;57:60–70.
7. Flickinger JC, Kondziolka D, Niranjan A, Maitz A, Voynov G, Lunsford LD. Acoustic neuroma radiosurgery with marginal tumor doses of 12 to 13Gy. Int J Radiat Oncol Biol Phys. 2004;60:225–30.
8. Hasegawa T, Kida Y, Kobayashi T, Yoshimoto M, Mori Y, Yoshida J. Long-term outcomes in patients with vestibular schwannomas treated using gamma knife surgery: 10-year follow up. J Neurosurg. 2005;102:10–6.
9. Hudgins WR, Antes KJ, Herbert MA, et al. Control of growth of vestibular schwannomas with low-dose gamma knife surgery. J Neurosurg. 2006;105:154–60.
10. Iwai Y, Yamanaka K, Shiotani M, Uyama T. Radiosurgery for acoustic neuromas: results of low-dose treatment. Neurosurgery. 2003;53:282–7.

11. Liu D, Xu D, Zhang Z, Zhang Y, Zheng L. Long-term outcomes after gamma knife surgery for vestibular schwannomas: a 10-year experience. J Neurosurg. 2006;105:149–53.
12. Lunsford LD, Niranjan A, Flickinger JC, Maitz A, Kondziolka D. Radiosurgery of vestibular schwannomas: summary of experience in 829 cases. J Neurosurg. 2005;102:195–9.
13. McEvoy AW, Kitchen ND. Rapid enlargement of a vestibular schwannoma following gamma knife treatment. Minim Invasive Neurosurg. 2003;46:254–6.
14. Mendenhall WM, Friedman WA, Buatti JM, Bova FJ. Preliminary results of linear accelerator radiosurgery for acoustic schwannomas. J Neurosurg. 1996;85:1013–9.
15. Miller RC, Foote RL, Coffey RL, et al. Decrease in cranial nerve complications after radiosurgery for acoustic neuromas: a prospective study of dose and volume. Int J Radiat Oncol Biol Phys. 1999;43:305–11.
16. Murphy ES, Barnett GH, Vogelbaum MA, et al. Long-term outcomes of gamma knife radiosurgery in patients with vestibular schwannomas. J Neurosurg. 2011;114:432–40.
17. Niranjan A, Mathieu D, Flickinger JC, Kondziolka D, Lunsford LD. Hearing preservation after intracanalicular vestibular schwannoma radiosurgery. Neurosurgery. 2008;63:1054–62.
18. Okunaga T, Matsuo T, Hayashi N, et al. Linear accelerator radiosurgery for vestibular schwannoma: measuring tumor volume changes on serial three-dimensional spoiled gradient-echo magnetic resonance images. J Neurosurg. 2005;103:53–8.
19. Sughrue ME, Yang I, Han SJ, et al. Non-audiofacial morbidity after gamma knife surgery for vestibular schwannoma. Neurosurg Focus. 2009;27:E4.
20. Yang I, Aranda D, Han SJ, et al. Hearing preservation after stereotactic radiosurgery for vestibular schwannoma: a systematic review. J Clin Neurosci. 2009;16:742–7.
21. Neuhaus O, Saleh A, van Oosterhout A, Siebler M. Cerebellar infarction after gamma knife radiosurgery of a vestibular schwannoma. Neurology. 2007;68:590.
22. Pollack AG, Marymont MH, Kalapurakal JA, Kepka A, Sathiaseelan V, Chandler JP. Acute neurological complications following gamma knife surgery for vestibular schwannoma. J Neurosurg. 2005;103:546–51.
23. Pollock BE. Management of vestibular schwannomas that enlarge after stereotactic radiosurgery: treatment recommendations based on a 15-year experience. Neurosurgery. 2006;58:241–8.
24. Demetriades AK, Saunders N, Rose P, et al. Malignant transformation of acoustic neuroma/vestibular schwannoma 10 years after gamma knife stereotactic radiosurgery. Skull Base. 2010;20:381–7.
25. Hanabusa K, Morikawa A, Murata T, Taki W. Acoustic neuroma with malignant transformation. J Neurosurg. 2001;95:518–21.
26. Husseini ST, Piccirillo E, Sanna M. On "malignant transformation of acoustic neuroma/vestibular schwannoma 10 years after gamma knife stereotactic radiosurgery" (skull base 2010;20:381–388). Skull Base. 2011;21:135–8.
27. Markou K, Eimer S, Perret C, et al. Unique case of malignant transformation of a vestibular schwannoma after fractionated radiotherapy. Am J Otolaryngol. 2012;33:168–73.
28. Rowe J, Grainger A, Walton L, Silcocks P, Radatz M, Kemeny A. Risk of malignancy after gamma knife stereotactic radiosurgery. Neurosurgery. 2007;60:60–6.
29. Schmitt WR, Carlson ML, Giannini C, Driscoll CL, Link MJ. Radiation-induced sarcoma in a large vestibular schwannoma following stereotactic radiosurgery: case report. Neurosurgery. 2011;68:E840–6.
30. Shin M, Ueki K, Kurita H, Kirino T. Malignant transformation of a vestibular schwannoma after gamma knife radiosurgery. Lancet. 2002;360:309–10.
31. Tanbouzi Husseini S, Piccirillo E, Taibah A, Paties CT, Rizzoli R, Sanna M. Malignancy in vestibular schwannoma after stereotactic radiotherapy: a case report and review of the literature. Laryngoscope. 2011;121:923–8.
32. Wilkinson JS, Reid H, Armstrong GR. Malignant transformation of a recurrent vestibular schwannoma. J Clin Pathol. 2004;57:109–10.
33. Yanamadala V, Williamson R, Fusco DJ, Eschbacher J, Weisskopf P, Porter R. Malignant transformation of a vestibular schwannoma after gamma knife radiosurgery: case report and review of the literature. World Neurosurg. 2013;79(593):e1–8. https://doi.org/10.1016/j.wneu.2012.03.016.

34. Slattery WH III. Microsurgery after radiosurgery or radiotherapy for vestibular schwannomas. Otolaryngol Clin N Am. 2009;42:707–15.
35. Breshears JD, Osorio JA, Cheung SW, Barani IJ, Theodosopoulos PV. Surgery after primary radiation treatment for sporadic vestibular schwannomas: case series. Oper Neurosurg (Hagerstown). 2017;13:441–7.
36. Friedman RA, Berliner KI, Bassim M, Ursick J, Slattery WH 3rd, Schwartz MS. A paradigm shift in salvage surgery for radiated vestibular schwannoma. Otol Neurotol. 2011;32:1322–8.
37. Friedman RA, Brackmann DE, Hitselberger WE, Schwartz MS, Iqbal Z, Berliner KI. Surgical salvage after failed irradiation for vestibular schwannoma. Laryngoscope. 2005;115:1827–32.
38. Gerganov VM, Giordano M, Samii A, Samii M. Surgical treatment of patients with vestibular schwannomas after failed previous radiosurgery. J Neurosurg. 2012;116:713–20. https://doi.org/10.3171/2011.12.JNS111682.
39. Iwai Y, Ishibashi K, Nakanishi Y, Onishi Y, Nishijima S, Yamanaka K. Functional outcome of salvage surgery for vestibular schwannomas after failed gamma knife radiosurgery. World Neurosurg. 2016;90:385–90.
40. Iwai Y, Yamanaka K, Yamagata K, Yasui T. Surgery after radiosurgery for acoustic neuromas: surgical strategy and histological findings. Neurosurgery. 2007;60:ONS75–82.
41. Lee CC, Yen YS, Pan DH, et al. Delayed microsurgery for vestibular schwannoma after gamma knife radiosurgery. J Neurooncol. 2010;98:203–12.
42. Lee DJ, Westra WH, Staecker H, Long D, Niparko JK, Slattery WH III. Clinical and histopathologic features of recurrent vestibular schwannoma (acoustic neuroma) after stereotactic radiosurgery. Otol Neurotol. 2003;24:650–60.
43. Lee F, Linthicum F Jr, Hung G. Proliferation potential in recurrent acoustic schwannoma following gamma knife radiosurgery versus microsurgery. Laryngoscope. 2002;112:948–50.
44. Limb CJ, Long DM, Niparko JK. Acoustic neuromas after failed radiation therapy: challenges of surgical salvage. Laryngoscope. 2005;115:93–8.
45. Pollock BE, Lunsford LD, Kondziolka D, et al. Vestibular schwannoma management. Part II failed radiosurgery and the role of delayed microsurgery. J Neurosurg. 1998;89:949–55.
46. Shuto T, Inomori S, Matsunaga S, Fujino H. Microsurgery for vestibular schwannoma after gamma knife radiosurgery. Acta Neurochir. 2008;150:229–34.
47. Slattery WH III, Brackmann DE. Results of surgery following stereotactic irradiation for acoustic neuromas. Am J Otol. 1995;16:315–9.
48. Wise SC, Carlson ML, Tveiten QV, Driscoll CL, Myrseth E, Lund-Johansen M, Link MJ. Surgical salvage of recurrent vestibular schwannoma following prior stereotactic radiosurgery. Laryngoscope. 2016;126:2580–6.
49. Nagao O, Serizawa T, Higuchi Y, et al. Tumor shrinkage of vestibular schwannomas after gamma knife surgery: results after more than 5 years of follow-up. J Neurosurg. 2010;113:122–7.
50. Hong B, Krauss JK, Bremer M, Karstens JH, Heissler HE, Nakamura M. Vestibular schwannoma micrsurgery for recurrent tumors after radiation therapy or previous surgical resection. Otol Neurotol. 2013;35:171–81.
51. Kliesch S, Vogelgesang S, Benecke R, Horstmann GA, Schroeder HW. Malignant brain oedema after radiosurgery of a medium-sized vestibular schwannoma. Cent Eur Neurosurg. 2010;71:88–91.
52. Weber DC, Chan AW, Bussiere MR, et al. Proton beam radiosurgery for vestibular schwannoma: tumor control and cranial nerve toxicity. Neurosurgery. 2003;53:577–86.
53. Wiet RJ, Micco AG, Bauer GP. Complications of the gamma knife. Arch Otolaryngol Head Neck Surg. 1996;122:414–6.

Conclusions

Luciano Mastronardi, Alberto Campione,
and Takanori Fukushima

Vestibular schwannomas (VSs) are the most common neoplasms of the cerebello-pontine angle, making up 6–8% of all intracranial tumors. Up to 75% of patients are treated within 5 years after primary diagnosis. Reported annual growth rates of VSs vary between 0.3 and 4.8 mm; less than 4% of sporadic cases shrink spontaneously.

These findings could substantiate in some cases the "wait-and-scan" strategy for tumors with maximal extrameatal diameter <20 mm. However, it is very well known that during this period, many patients lose their hearing.

According to several authors, radiosurgery seems to be a safe and effective alternative treatment, but it does not cure VS. According to our experience, it seems to be reasonable to consider stereotactic radiosurgery for those patients harboring regrowth or progression of previously surgically treated VSs, who cannot be or do not want to be reoperated on.

VSs cause sensorineural hearing loss in up to 95% of affected individuals, but the mechanisms underlying this hearing loss are not always completely clear. Moreover, vertigo influences the quality of life regardless of the medical management strategy.

Total or "nearly total" tumor excision with the preservation of neurological function and quality of life is the goal of modern-day VS surgery. Recurrence rates after subtotal removal are three times higher than after complete removal.

In case of preoperative socially useful hearing and small tumor size (<2 cm), the chance of hearing preservation is higher than 50%. Anyway, if unilateral deafness could be acceptable to the patients, postoperative facial nerve paralysis is a devastating complication of VS surgery. Nowadays, facial nerve preservation is

L. Mastronardi (✉) · A. Campione
Department of Neurosurgery, San Filippo Neri Hospital—ASLRoma1, Rome, Italy
e-mail: mastro@tin.it; albertocampione@hotmail.it

T. Fukushima
Division of Neurosurgery, Duke University Medical Center, Carolina Neuroscience Institute, Raleigh, NC, USA
e-mail: Fukushima@carolinaneuroscience.com

© Springer Nature Switzerland AG 2019
L. Mastronardi et al. (eds.), *Advances in Vestibular Schwannoma Microneurosurgery*, https://doi.org/10.1007/978-3-030-03167-1

accomplished in more than 90% of cases. Moreover, attempts of preoperative prediction of the course of facial nerve by means of diffusion tensor imaging have proved successful and may yield even greater results in terms of facial nerve preservation in the future, allowing for a more accurate presurgical planning.

The origin of VS has always been a matter of debate. The vast majority of VSs originate from the inferior vestibular nerve; the incidence of involvement of this nerve increases proportionally to the tumor size. The inferior vestibular nerve as the nerve of origin of the tumor can be considered as one of the concurring factors determining poor functional outcome of cochlear nerve preservation and also probably accounts for the better hearing results reported in some case series with the retrosigmoid approach. Preoperative determination of the nerve of origin of VS may therefore serve as both a prognostic factor (in terms of hearing preservation probability) and a presurgical planning feature. The latest results from the literature show that the origin of VS can be predicted with good accuracy, and further studies are granted to establish the usefulness of such adjunct data in the setting of microsurgical intervention.

Experience of the medical team, interdisciplinarity, quality of the physician-patient relationship, use of modern technology, and the knowledge about the long-term results of observation and intervention influence treatment quality in patients with VS.

Intraoperative neurophysiological monitoring has become an integral part of VS surgery. We have described several techniques of intraoperative neurophysiological monitoring, identified the clinical impact of certain pathognomonic patterns on postoperative outcomes of facial nerve function and hearing preservation, and highlighted the role of postoperative medications in improving delayed cranial nerve dysfunction in the different reported series. Recent advances in electrophysiological technology have considerably contributed to improvement in functional outcome of VS surgery in terms of hearing preservation and facial nerve paresis. Perioperative intravenous nimodipine and intraoperative diluted papaverine may be valuable adjuncts to surgery.

Quality of life is defined by symptoms caused primarily by the tumor itself and only secondarily by the medical interventions. Treatment should be directed at the preservation of the patient's quality of life from the beginning. Results of medical treatment should be superior to the natural course of the disease.

Our results and those reported in the international literature show low rates of morbidity and mortality. Microneurosurgical techniques are helpful for total resection of acoustic neuroma and anatomical preservation of facial and cochlear nerves. However, notwithstanding that complete removal is the main target of surgery, adoption of subtotal or "nearly total" removal strategies in selected cases with severe adhesions to facial nerve and/or brainstem can improve postoperative results and reduce duration and risks of surgery.

MIX
Papier | Fördert
gute Waldnutzung
FSC® C083411

Zeitfracht Medien GmbH
Ferdinand-Jühlke-Straße 7
99095 Erfurt, Deutschland
produktsicherheit@kolibri360.de